INTRODUCTION

Nobody can deny the importance of arithmetic in the advancement of individual and societal well-being. Numbers are an abstract concept, but combined with operations, they can express things language, art and other media can't. Highly complex phenomena are usually expressed in a short and simple mathematical formulas. Learning and practicing math is essential for all. Society and human condition will improve tremendously if all of us are adequately armored with math. It has been repeatedly shown that individuals who practice mental math, puzzles, and games maintain a sharper mind.

This book offers a brand new idea for a number puzzle. It is simple, yet challenging. One can feel that working with the puzzles in this book is greatly rewarding.

The main idea implemented in this book is "reverse addition". One can readily add numbers like (5 + 6) to get 11. But what are two different numbers can add up to 11? When solving that question, there are several answers.

This book has 400 puzzles, each of which takes the shape of a pentagon. Five numbers are given on the 5 sides of the pentagon. There are 5 empty circles given on the vertices of the pentagon. The goal of each puzzle is to find what numbers belong inside of those 5 circles. To do this, one must use "reverse addition." By adding the two numbers of adjacent circles, one must obtain the number between them. The players can use educated guesses, trial and error, and problem solving techniques. There are no answers given to the puzzles because each puzzle can be easily checked with simple addition. If a puzzle is solved correctly, it will be self-consistent.

The puzzles are well fit for all ages above 8, and they progress from easy to difficult. This book can be used in the classroom as an engaging exercise in mathematics and problem solving.

This is a continuation of our efforts in this field. This work follows and draws from our previous ideas in the book "Addition Addiction."

INSTRUCTION

This book is a tremendous exercise in REVERSE ADDITION. In each of the 400 puzzles, you are given a pentagon with five integers written on its five sides. Each pentagon has five vertices (nodes) with small empty circles.

The goal of these puzzles is to find five integers to write in the five empty circles of the pentagon, such that when the numbers in each pair of adjacent circles are added, the sum of the numbers will be the integer given on the connecting line between them.

TIPS

To solve for the missing integers and figure out the puzzle, you can use trial and error, insight, and problem solving techniques. Start from the weakest link in the pentagon, which is a side with the smallest of the five given numbers. This way you do not have as many combinations to consider as when you deal with a larger number on the pentagon.

Time yourself to track your progression, and solve each puzzle without an algebraic formula for an engaging experience!

Keep tuned for news on more new and related exciting books that are coming soon.

Puzzle 1

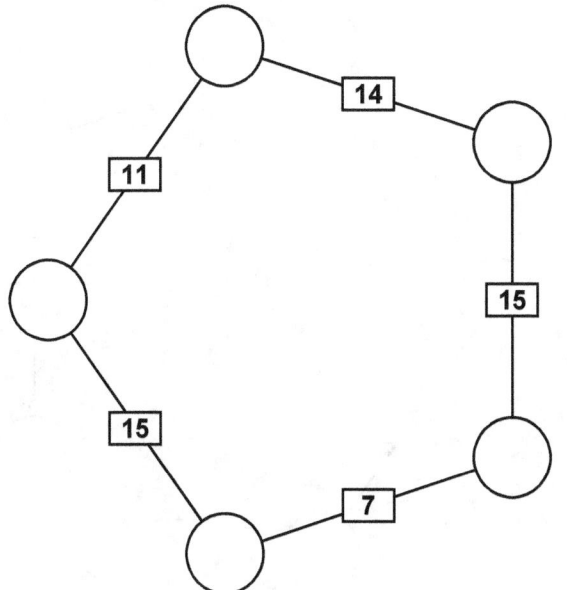

Puzzle 2

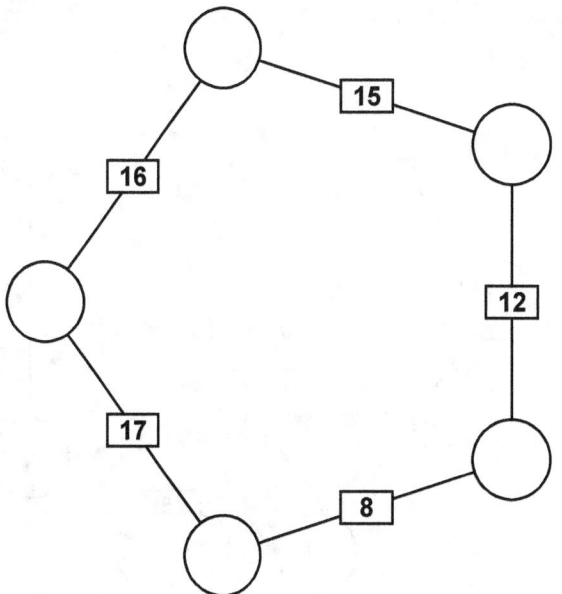

Puzzle 3

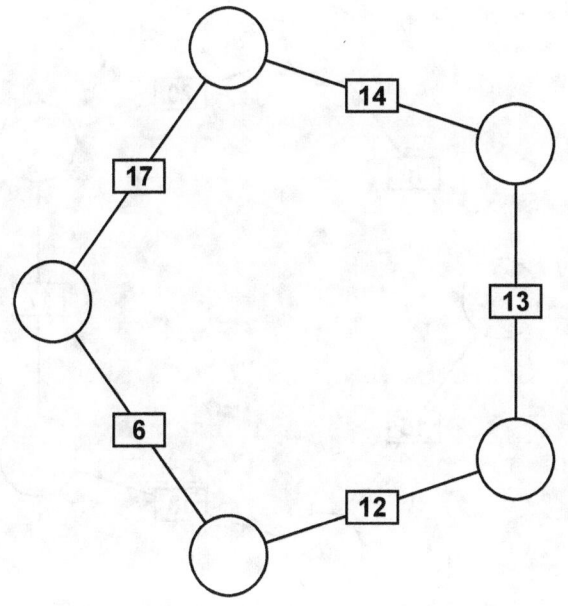

Puzzle 4

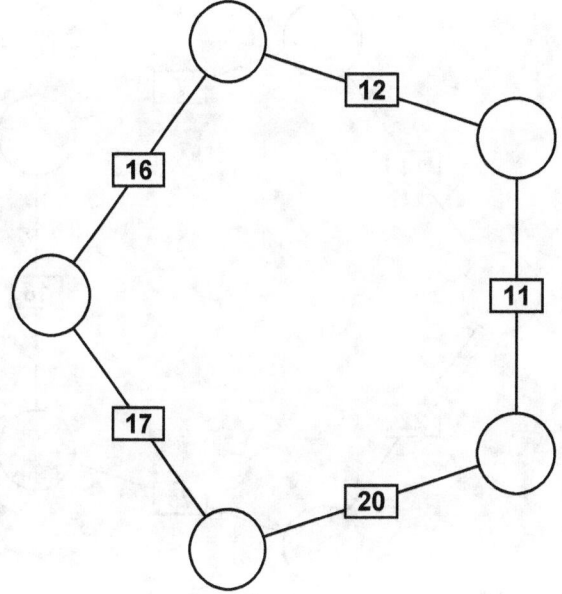

Puzzle 5

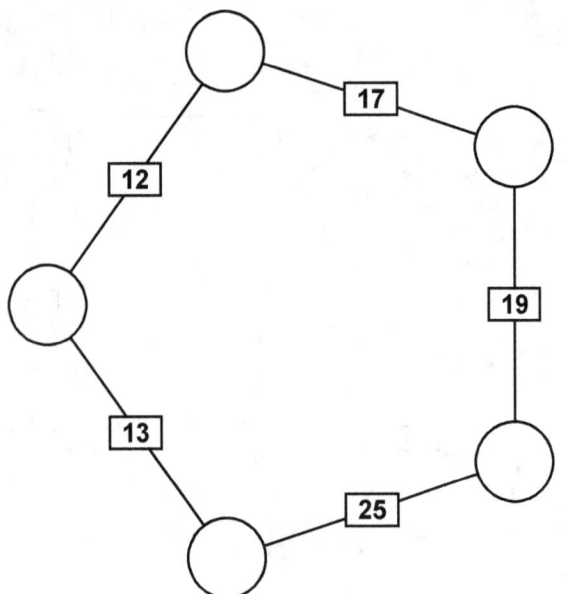

Puzzle 6

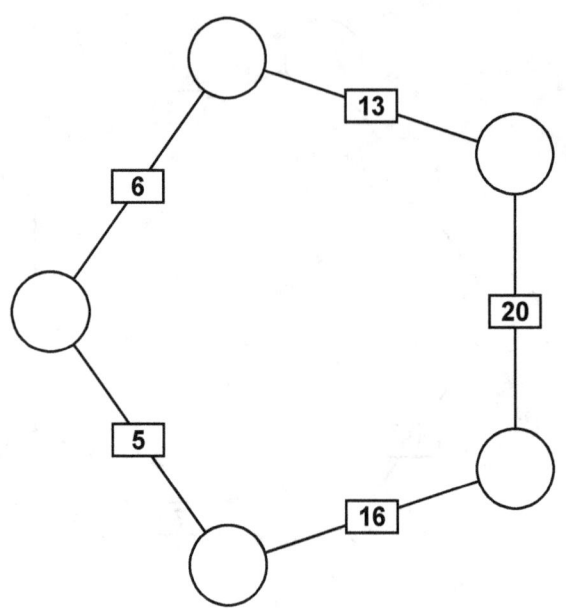

Puzzle 7

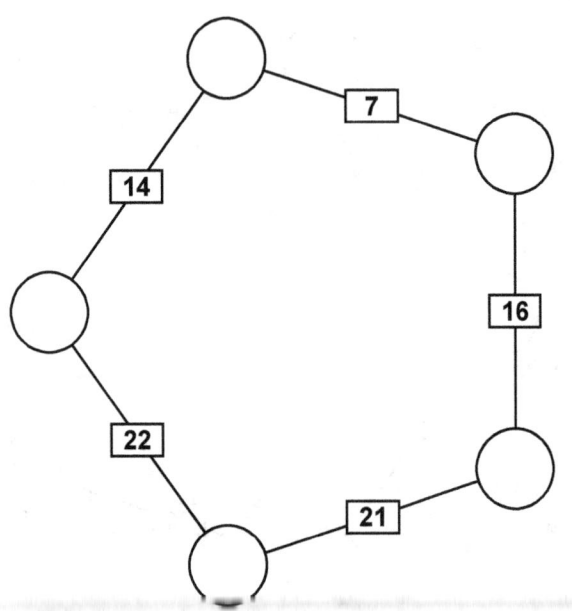

Puzzle 8

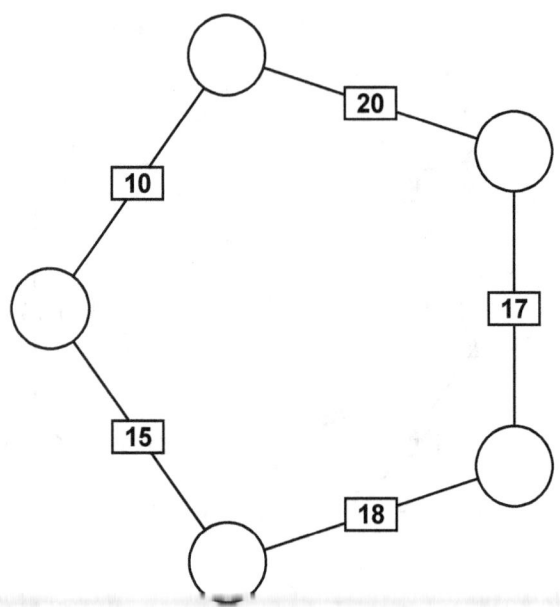

Puzzle 9

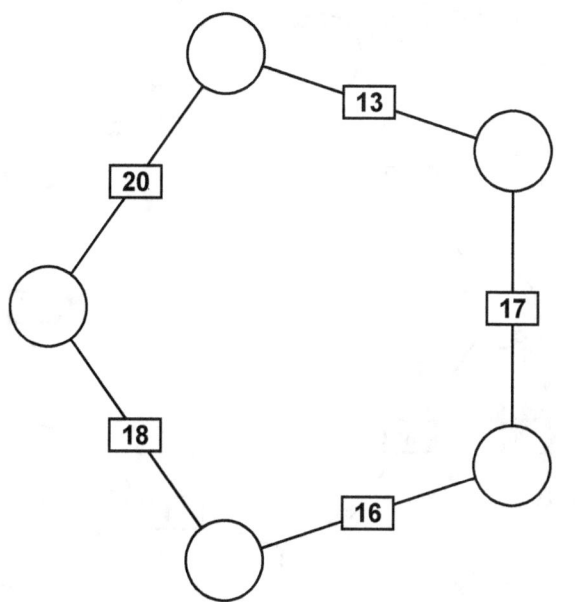

Puzzle 10

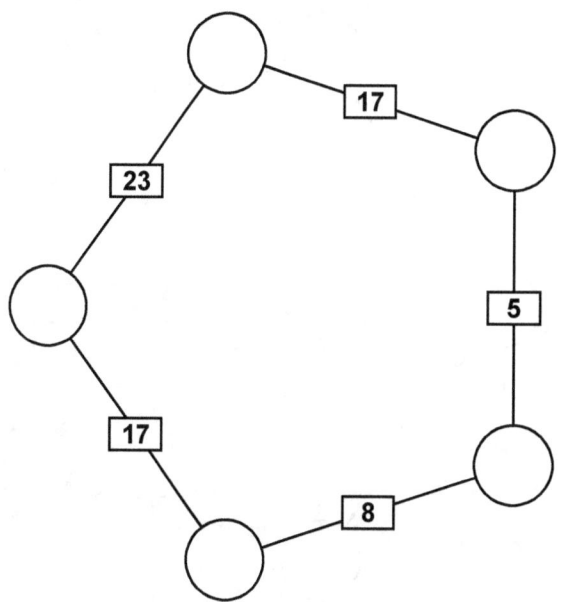

Puzzle 11

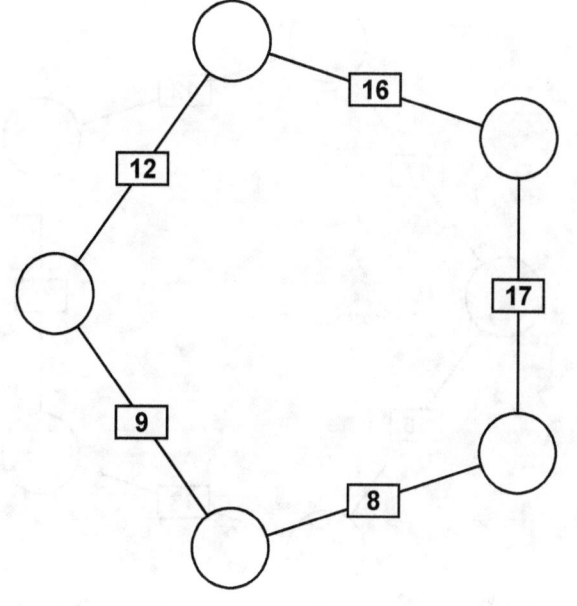

Puzzle 12

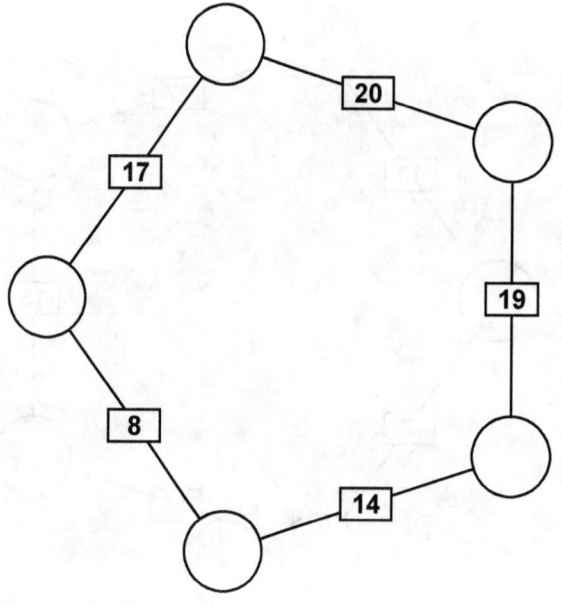

Puzzle 13

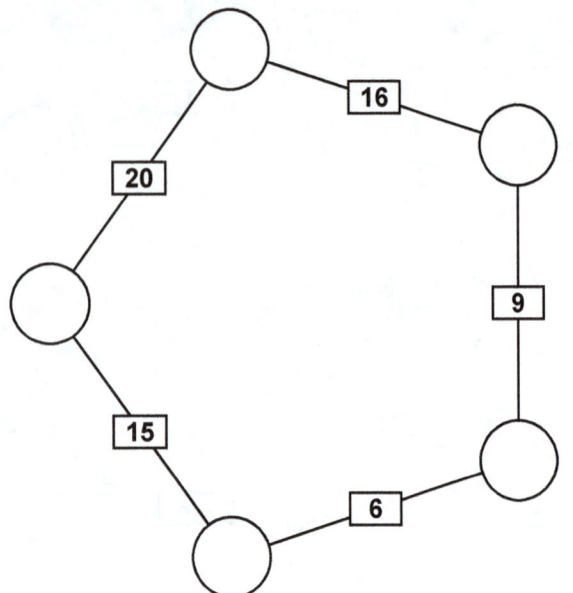

Puzzle 14

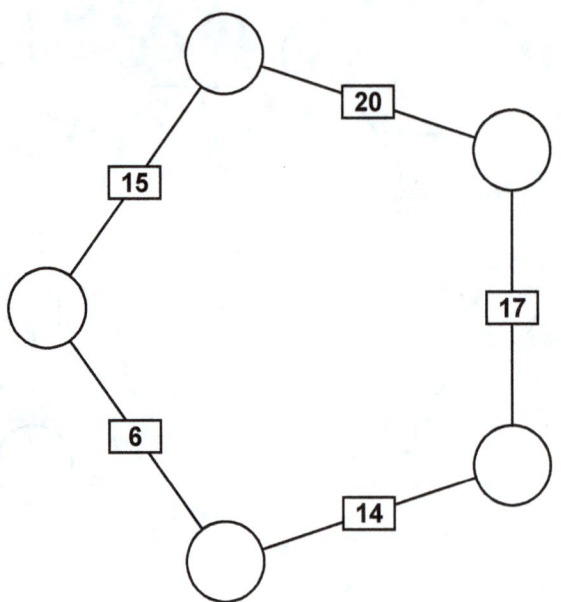

Puzzle 15

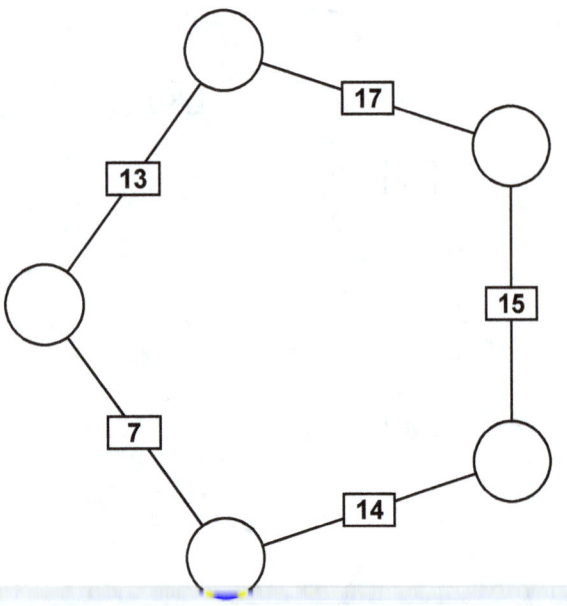

Puzzle 16

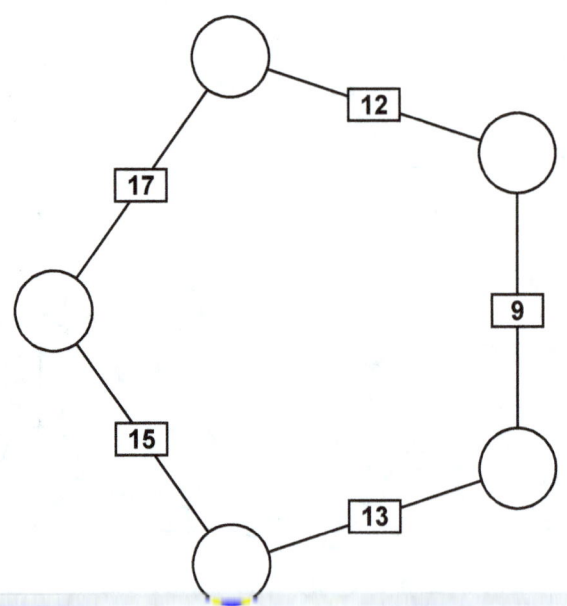

Puzzle 17

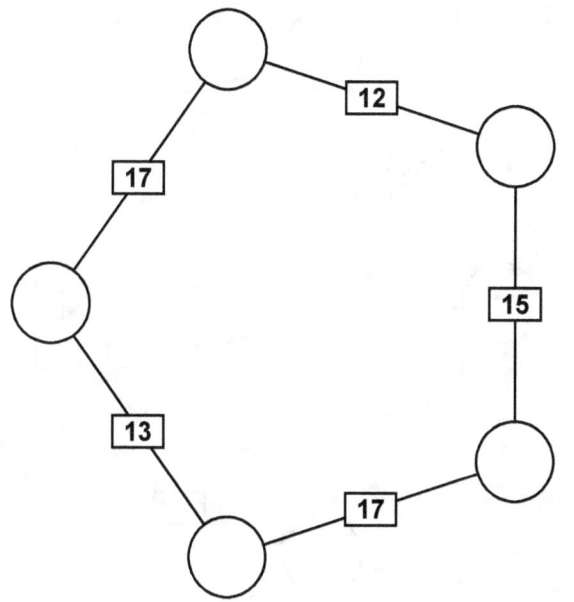

Puzzle 18

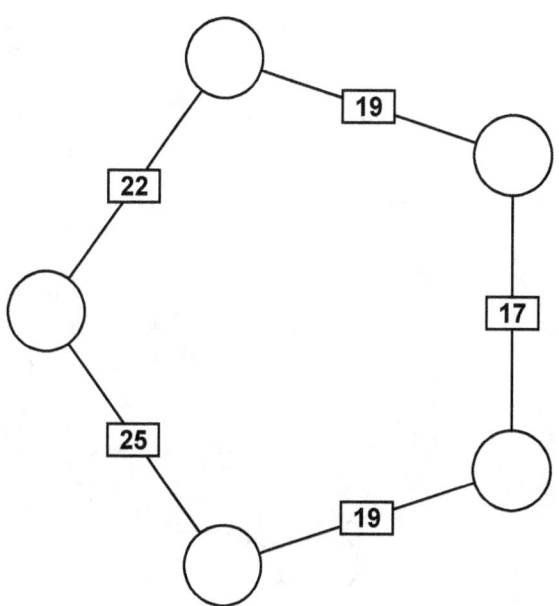

Puzzle 19

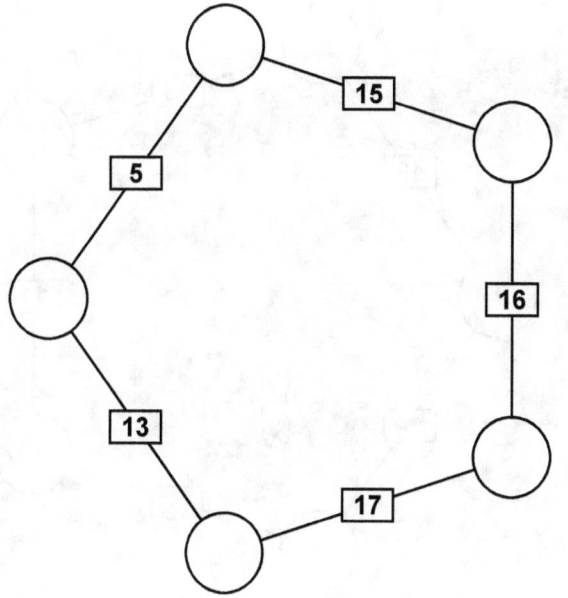

Puzzle 20

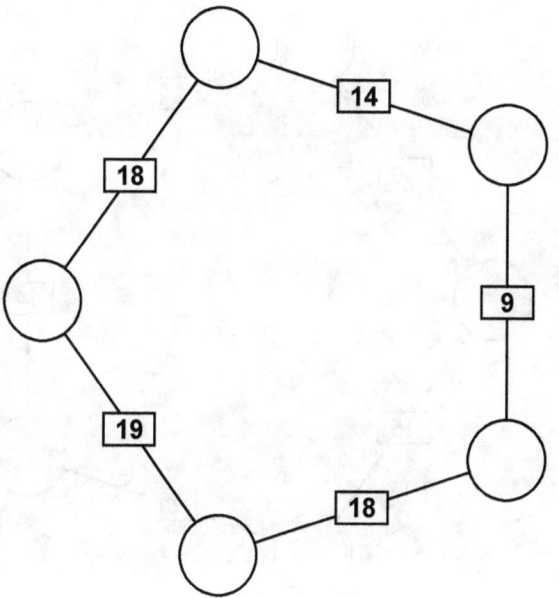

Puzzle 21

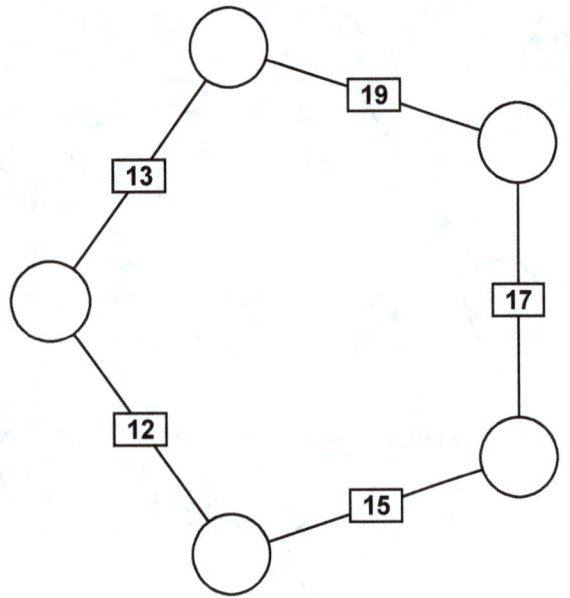

Puzzle 22

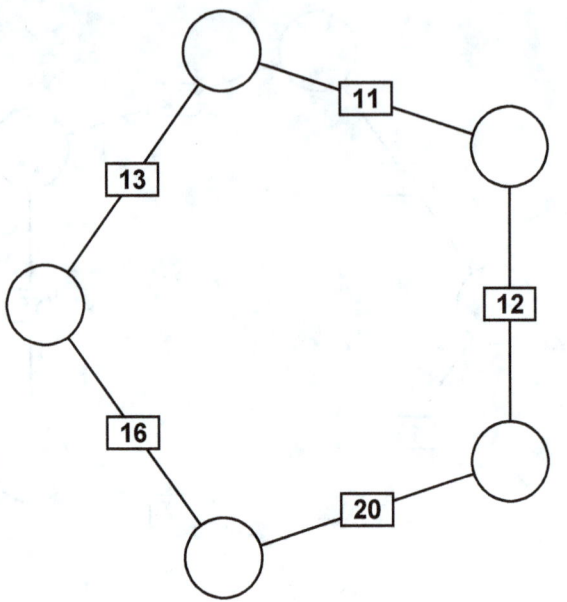

Puzzle 23

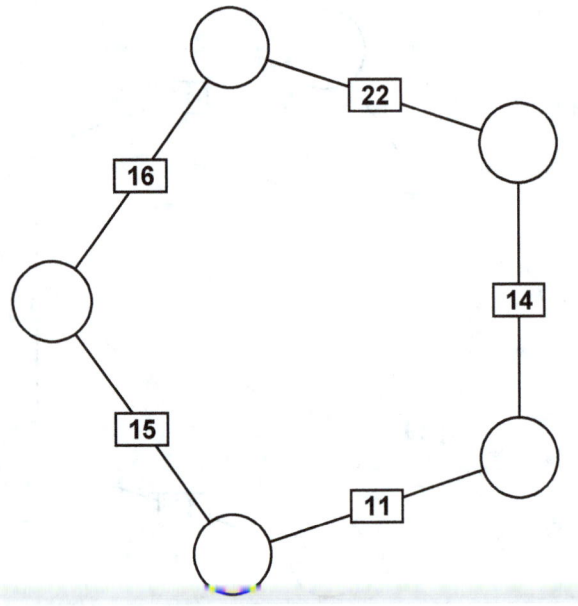

Puzzle 24

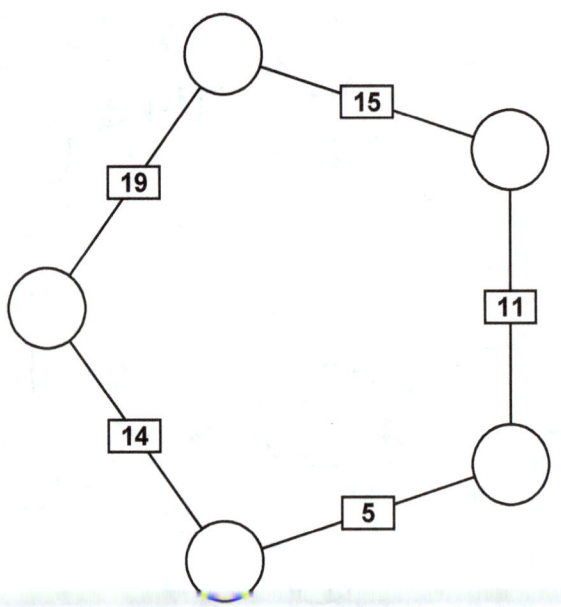

Puzzle 25

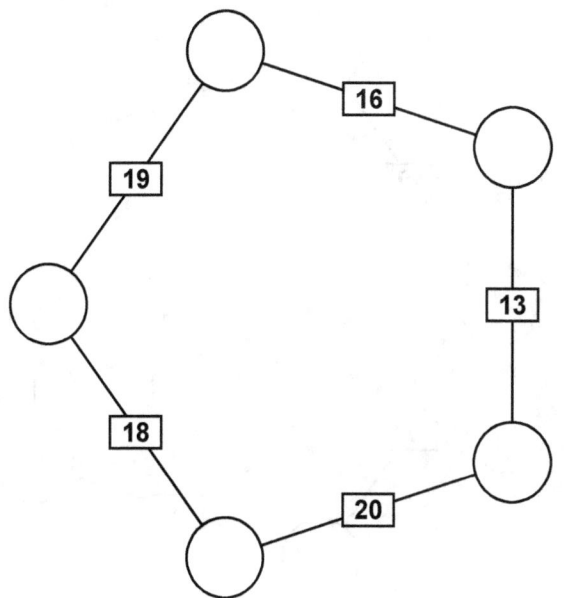

Puzzle 26

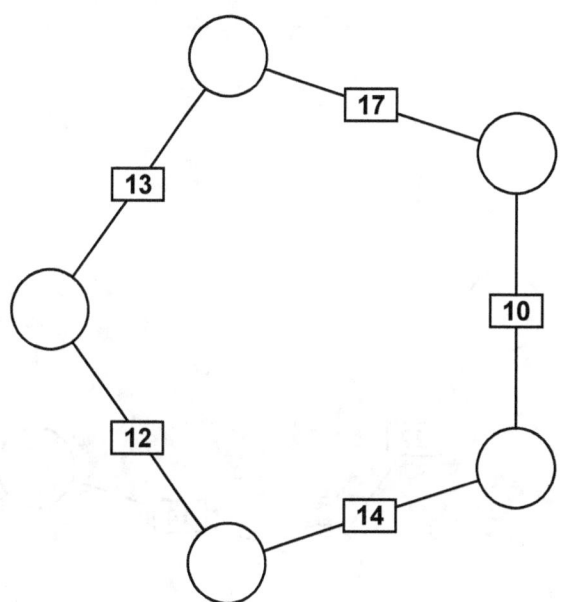

Puzzle 27

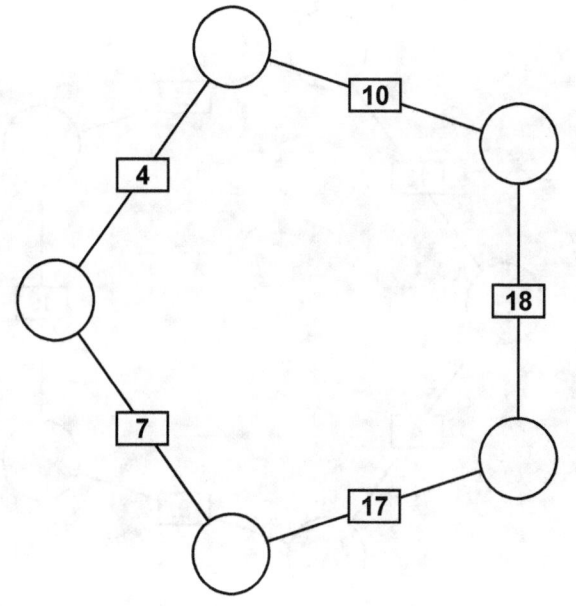

Puzzle 28

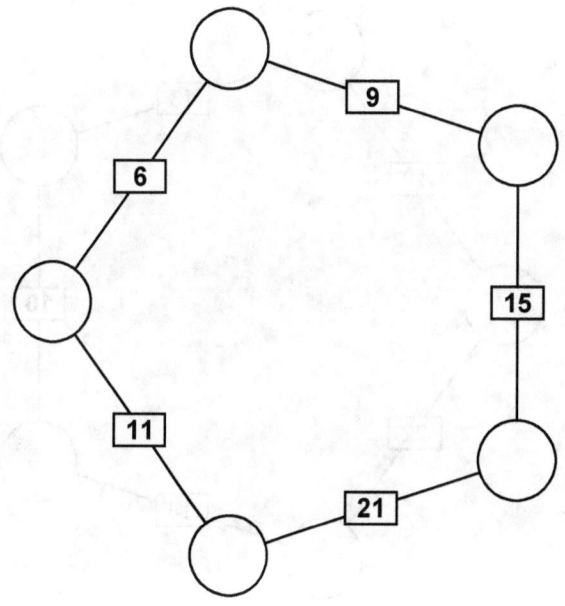

Puzzle 29

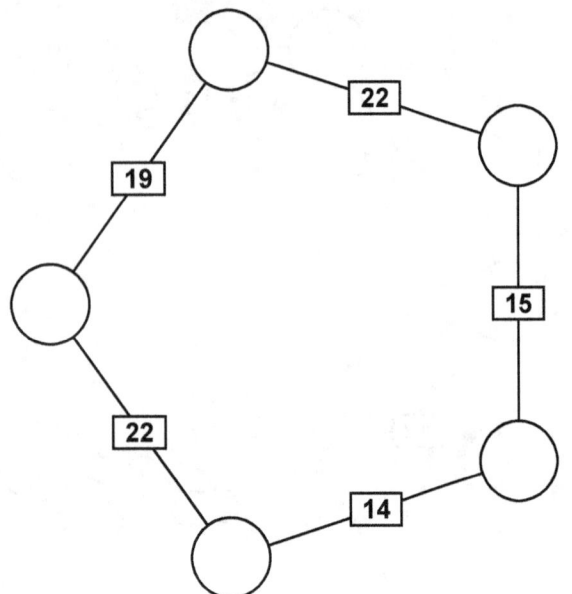

Puzzle 30

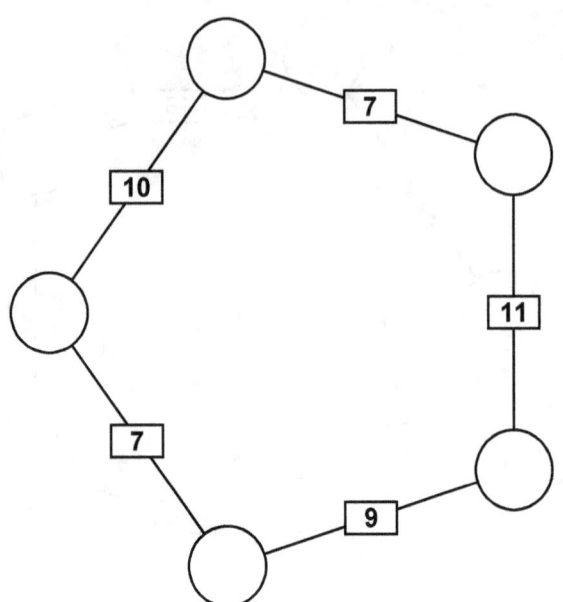

Puzzle 31

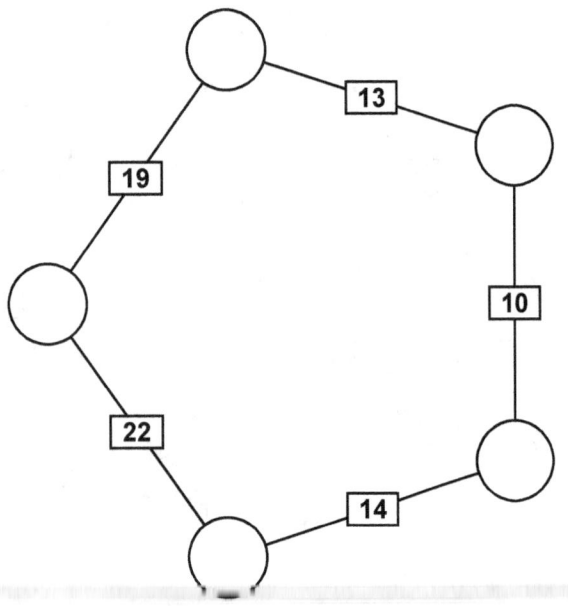

Puzzle 32

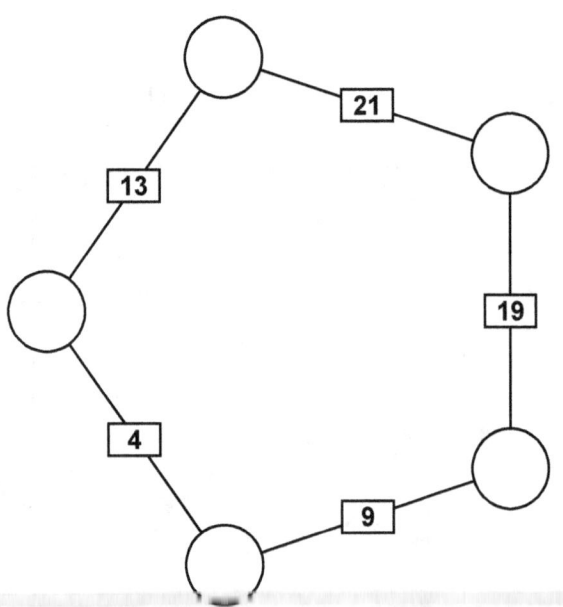

Puzzle 33

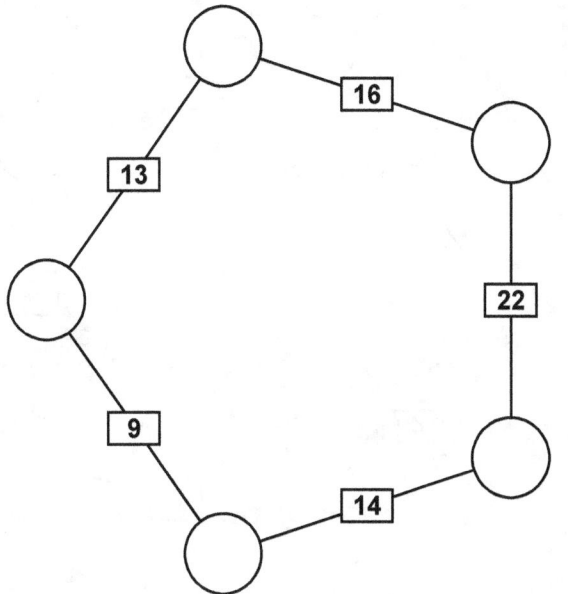

Puzzle 34

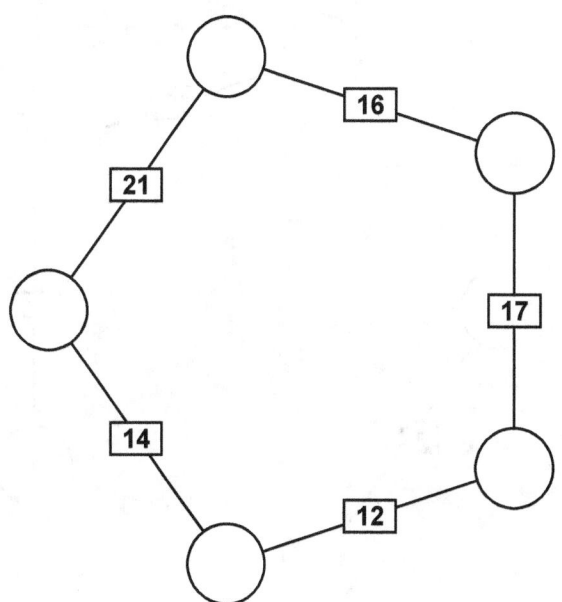

Puzzle 35

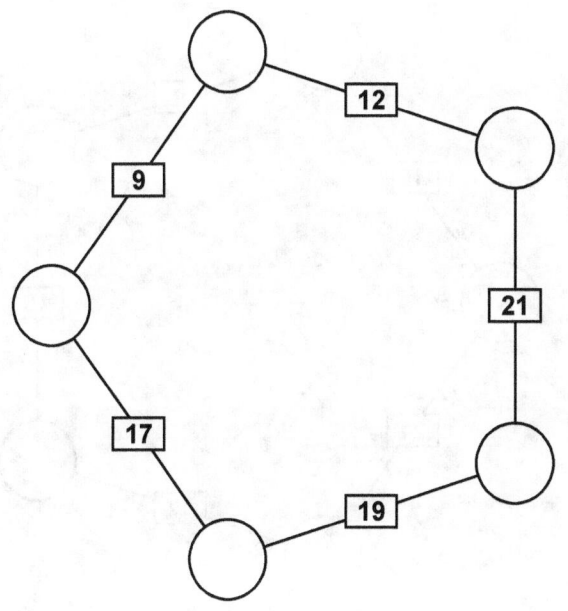

Puzzle 36

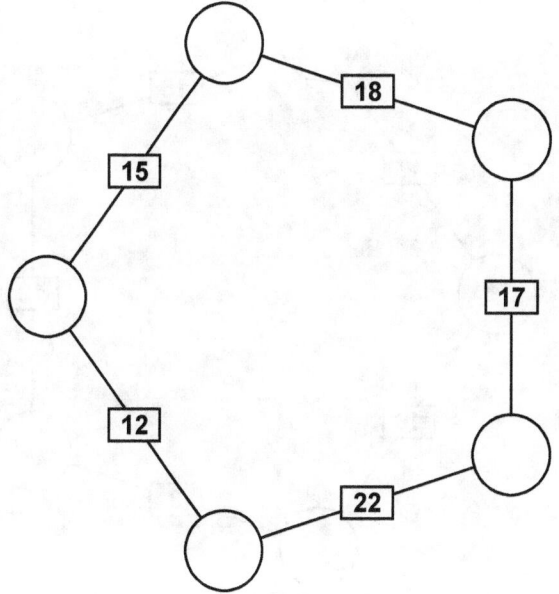

Puzzle 37

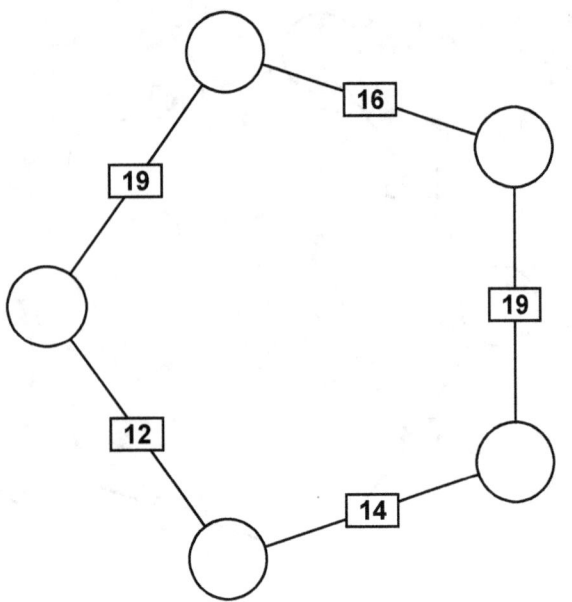

Puzzle 38

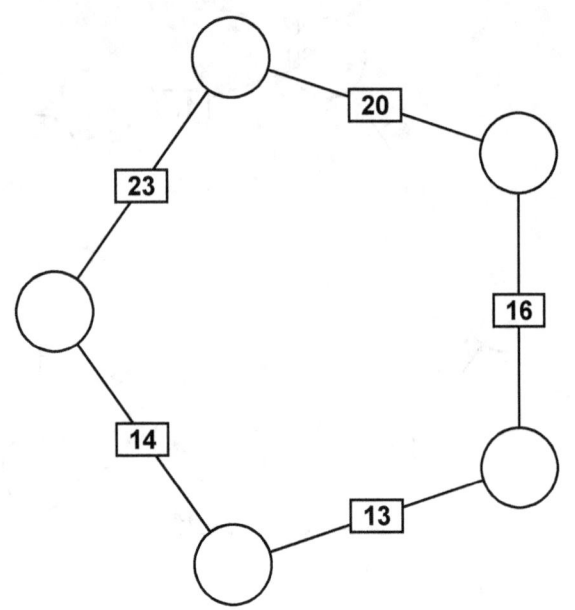

Puzzle 39

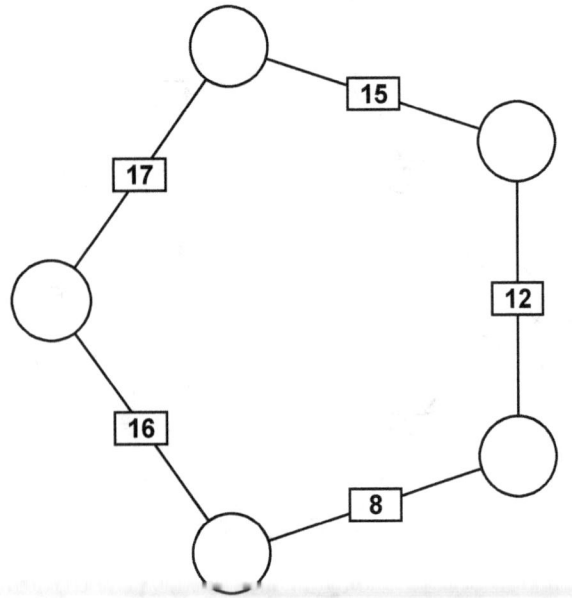

Puzzle 40

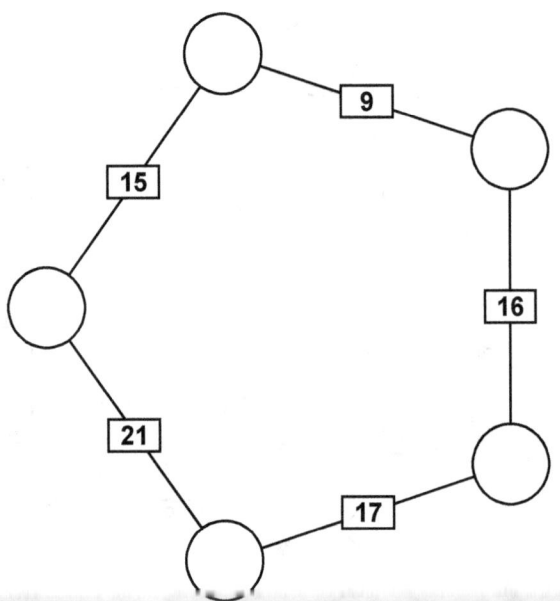

Puzzle 41

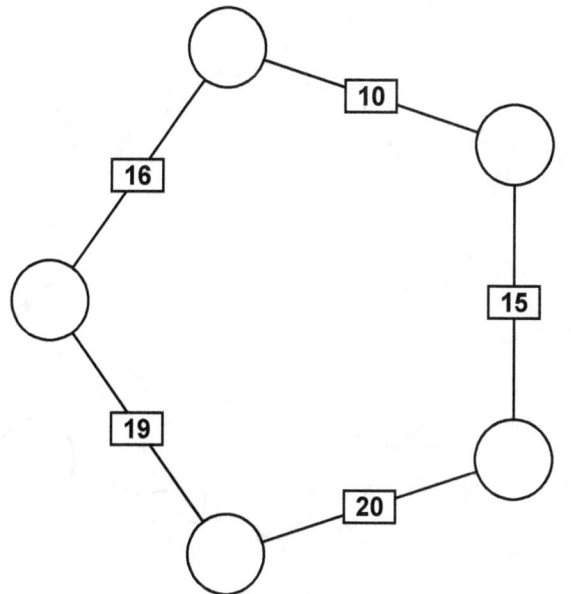

Puzzle 42

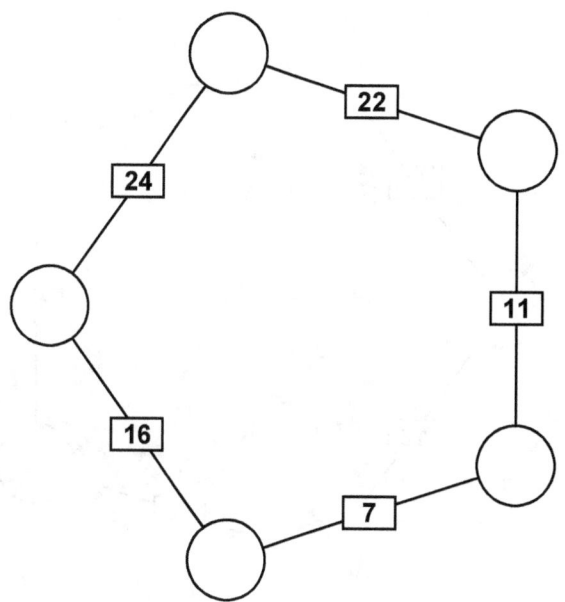

Puzzle 43

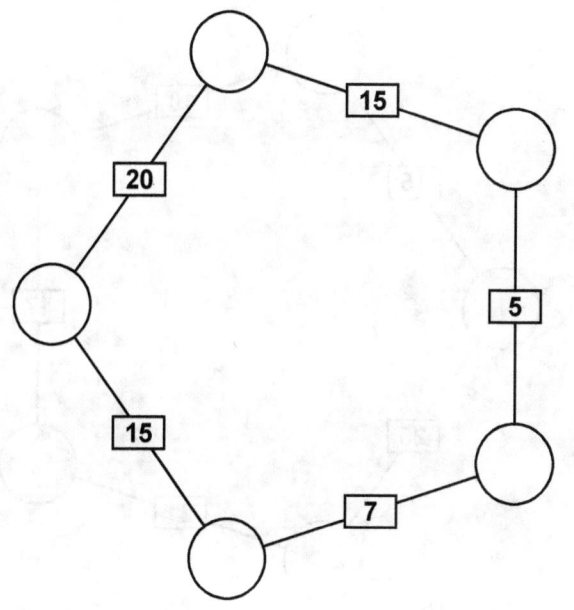

Puzzle 44

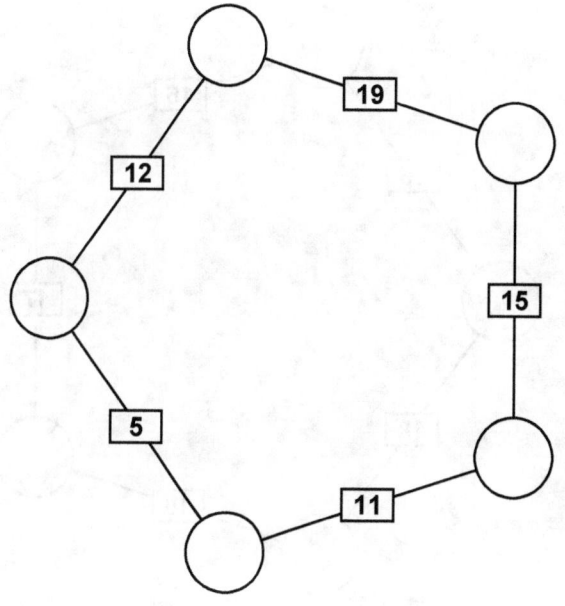

Puzzle 45

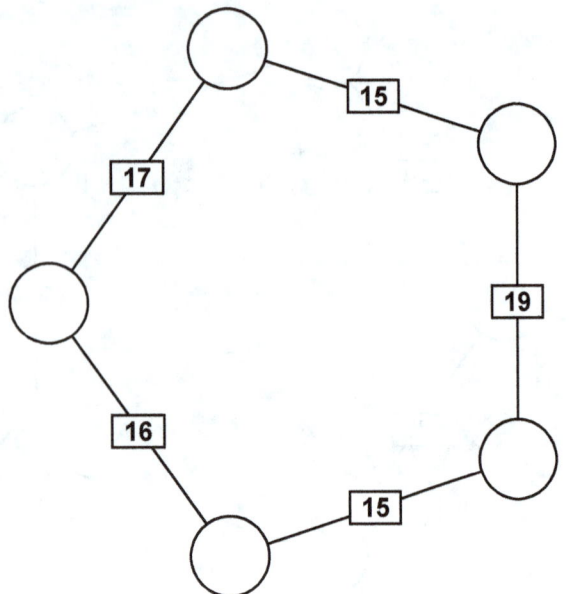

Puzzle 46

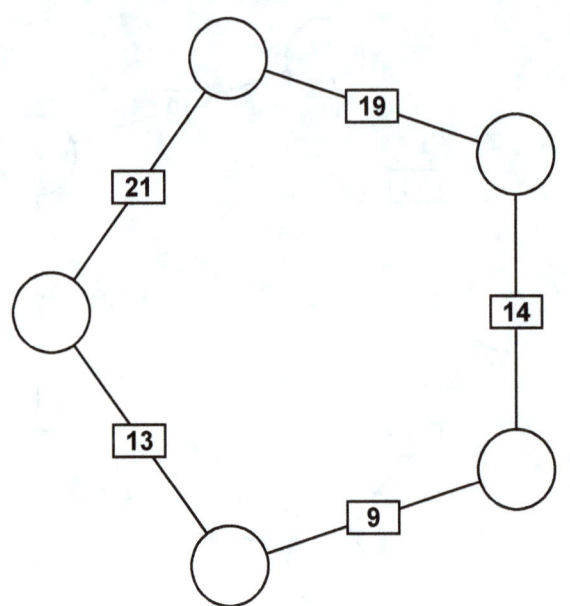

Puzzle 47

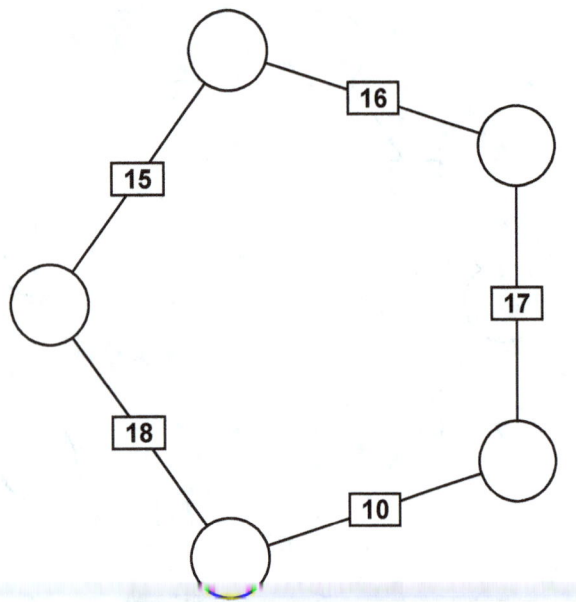

Puzzle 48

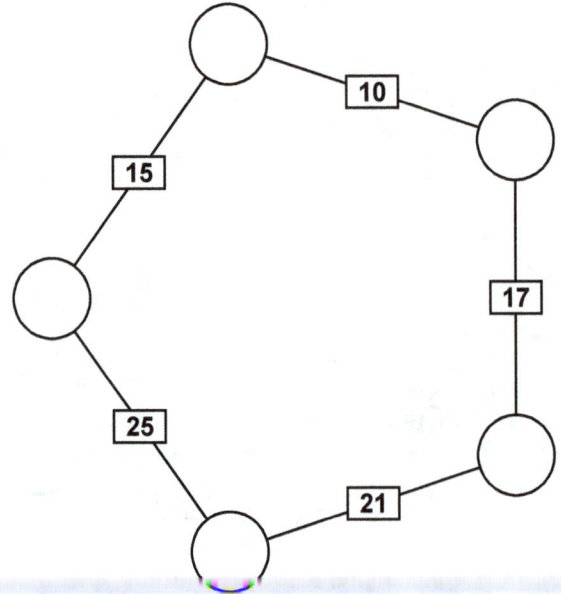

Puzzle 49

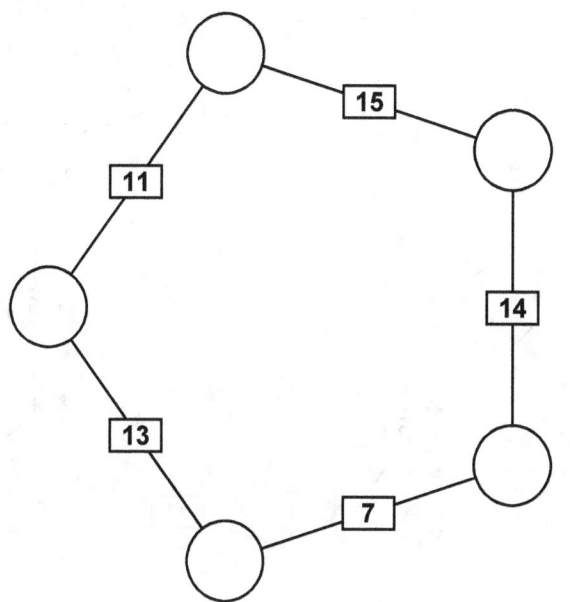

Puzzle 50

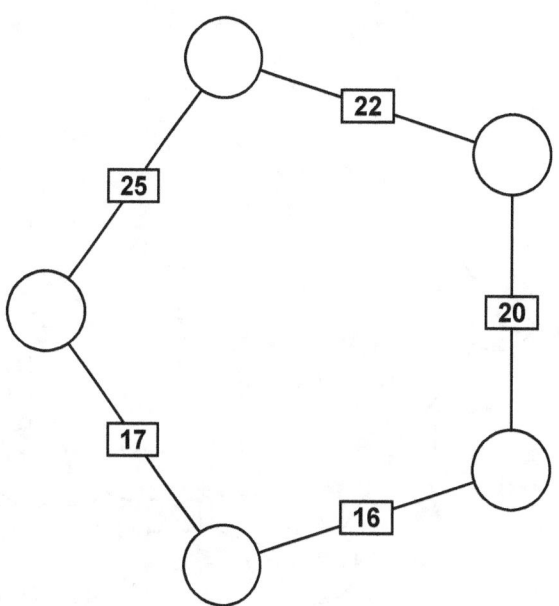

Puzzle 51

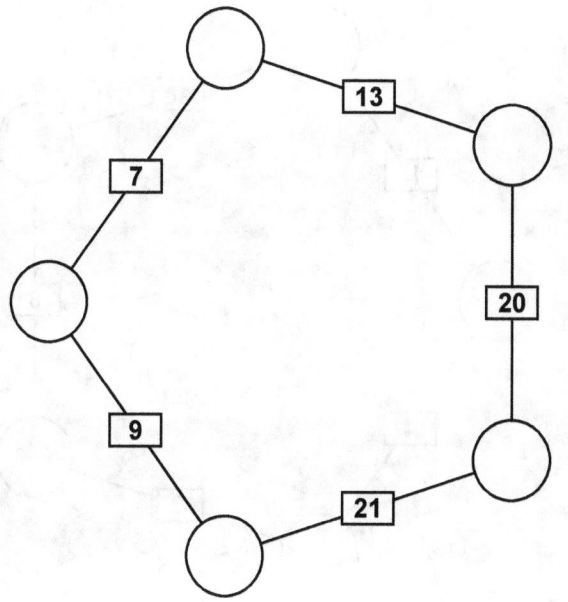

Puzzle 52

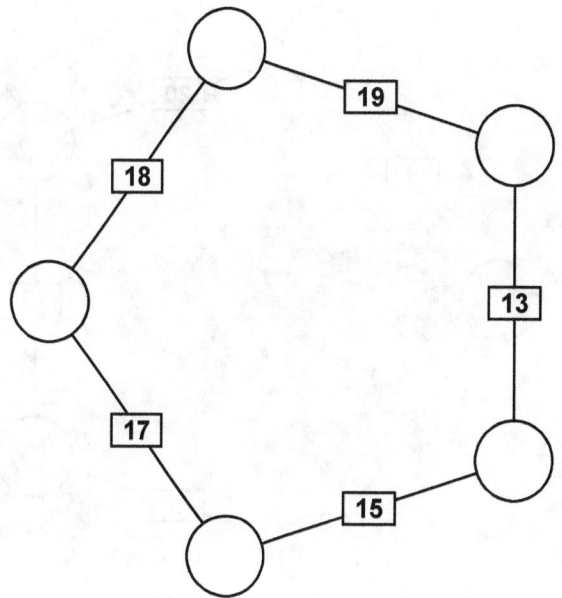

Puzzle 53

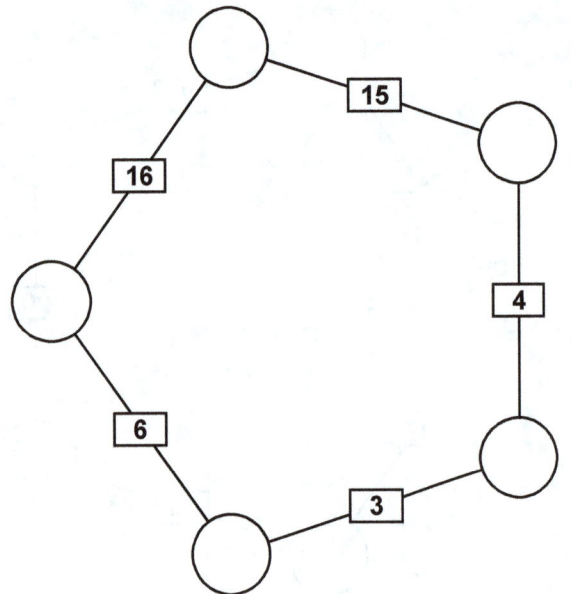

Puzzle 54

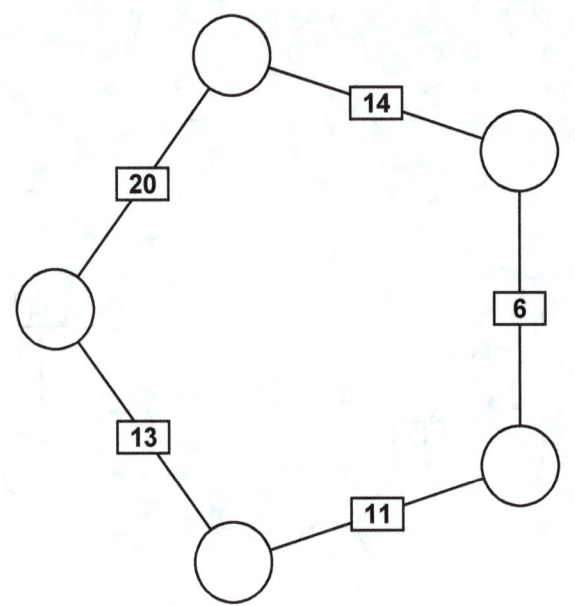

Puzzle 55

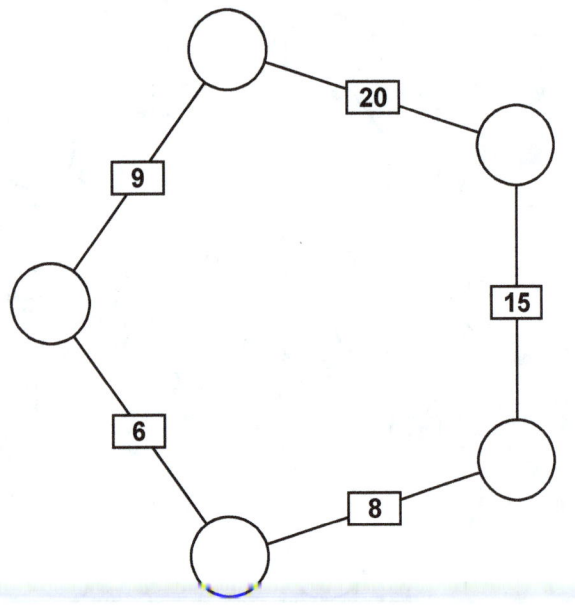

Puzzle 56

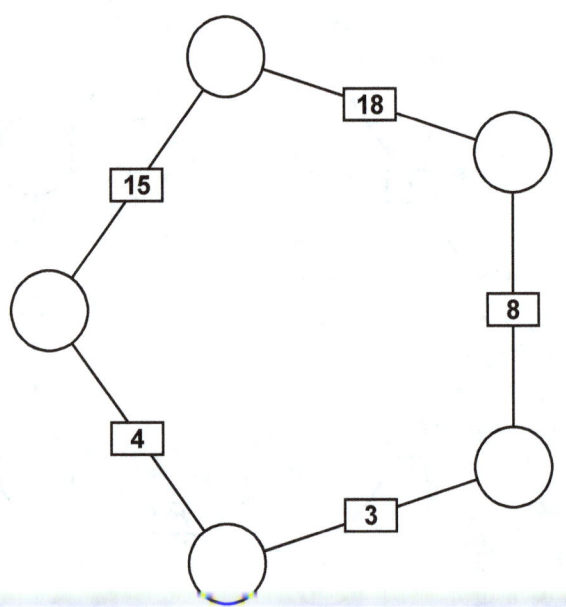

Puzzle 57

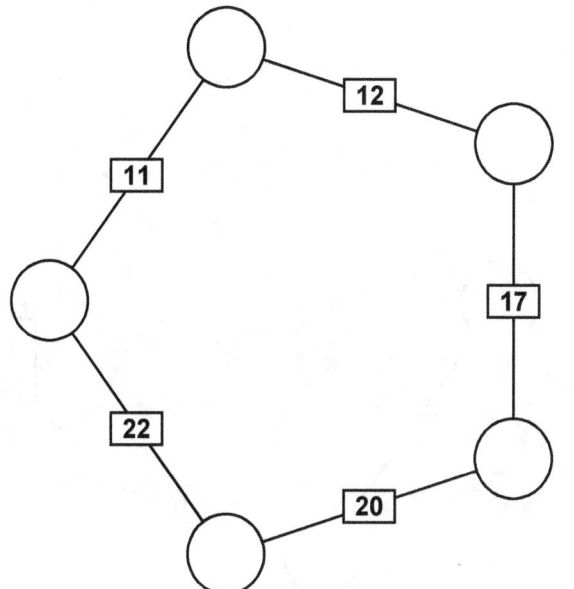

Puzzle 58

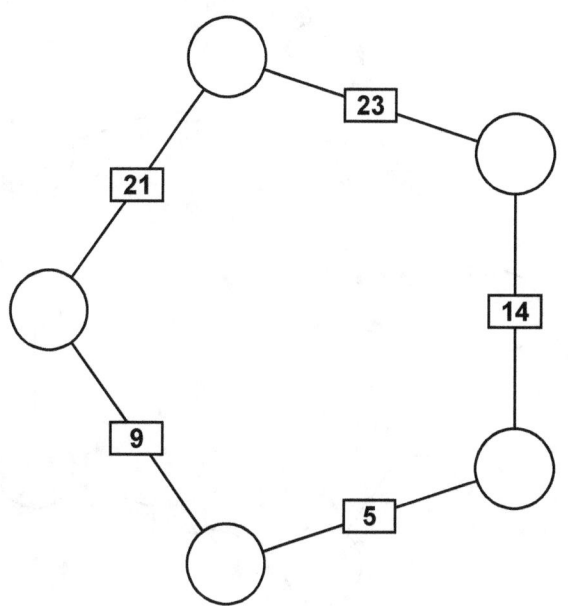

Puzzle 59

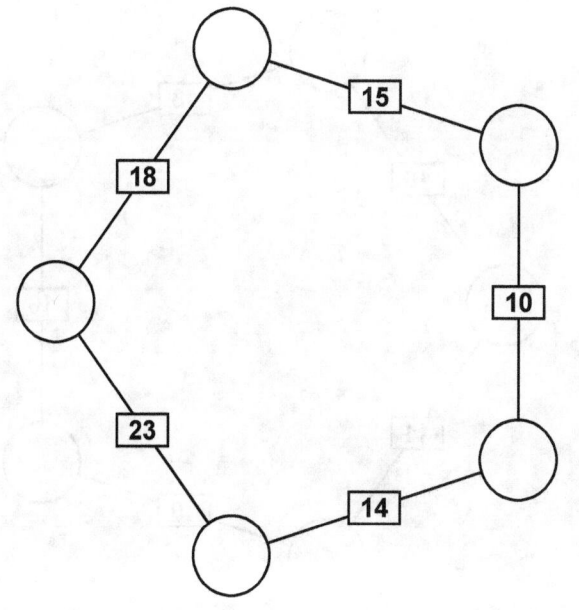

Puzzle 60

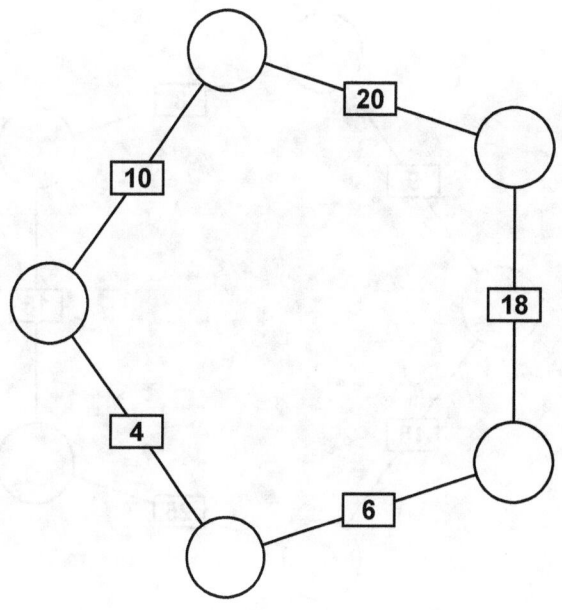

Puzzle 61

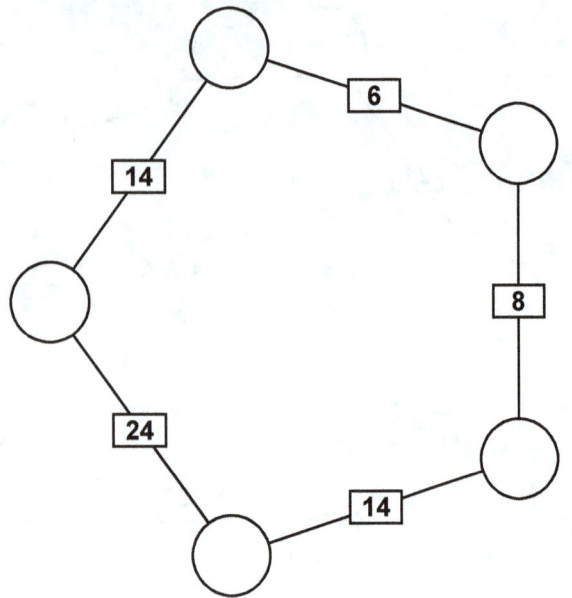

Puzzle 62

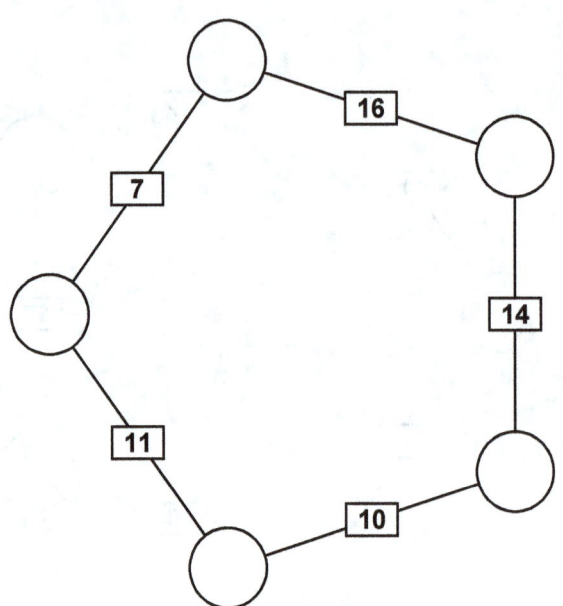

Puzzle 63

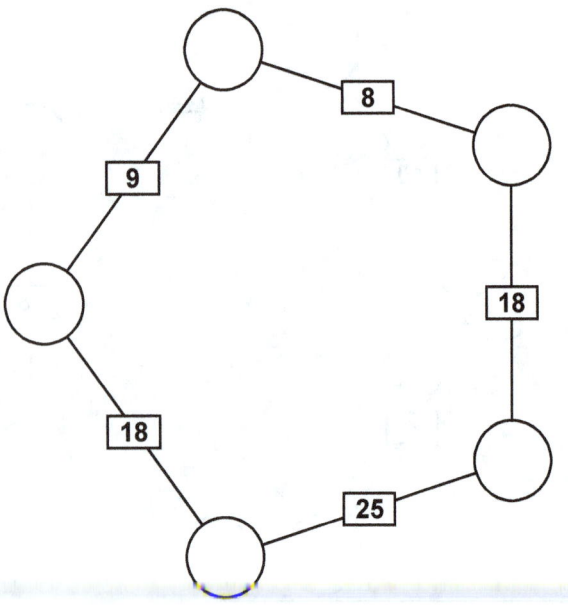

Puzzle 64

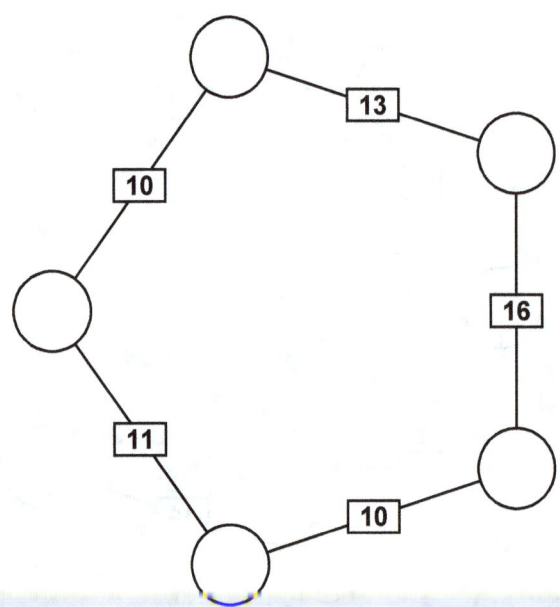

Puzzle 65

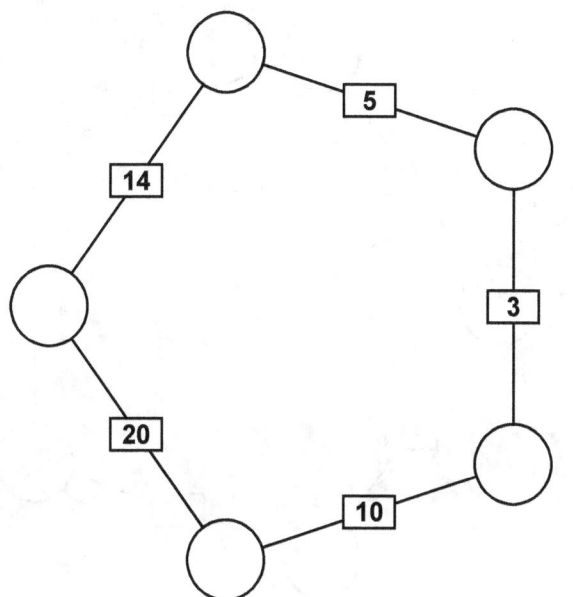

Puzzle 66

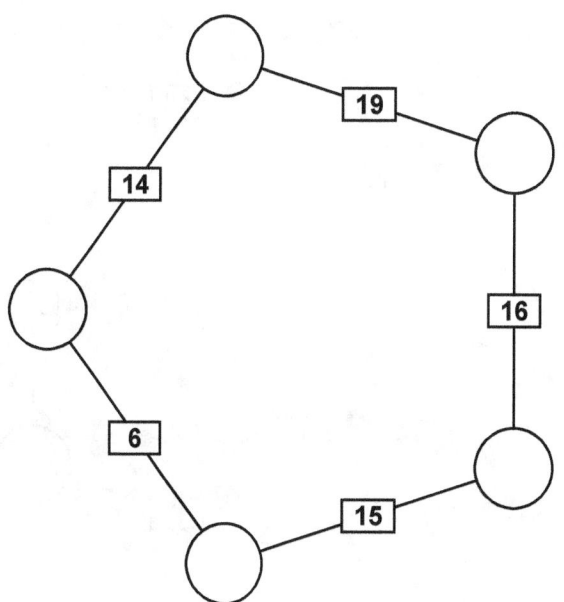

Puzzle 67

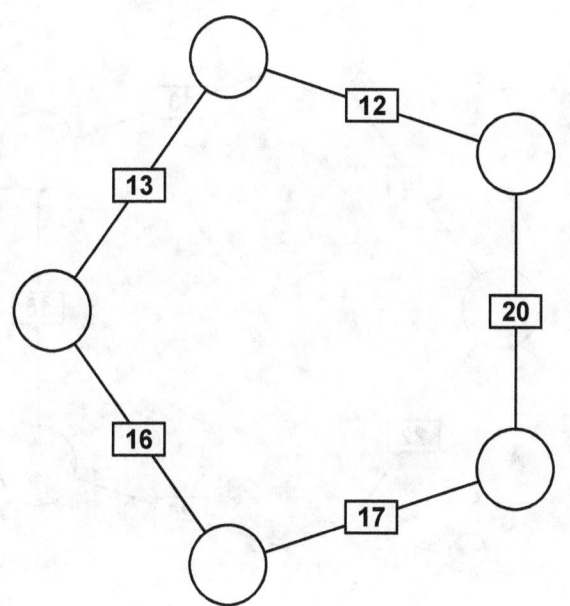

Puzzle 68

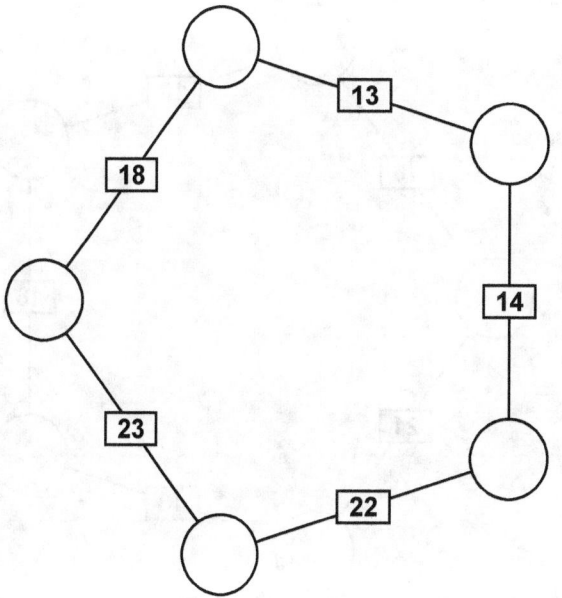

Puzzle 69

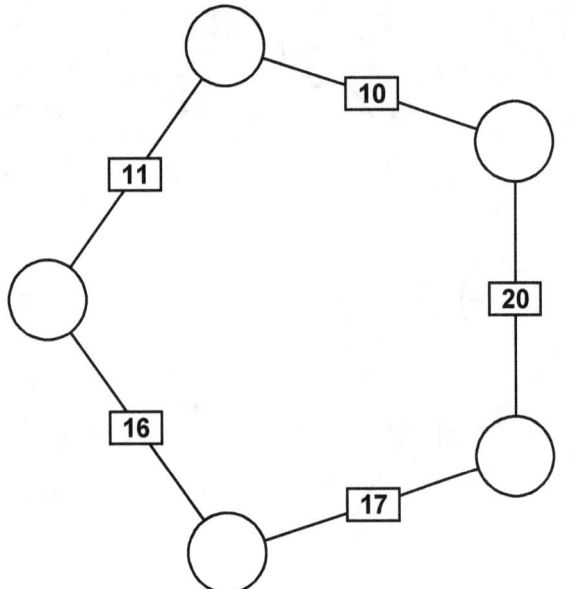

Puzzle 70

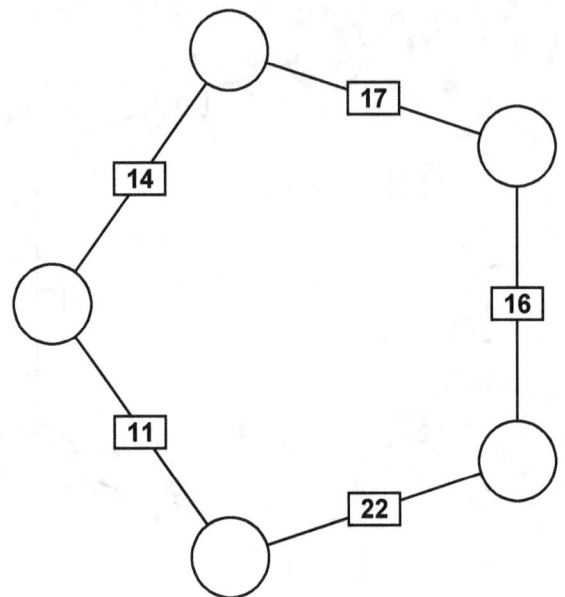

Puzzle 71

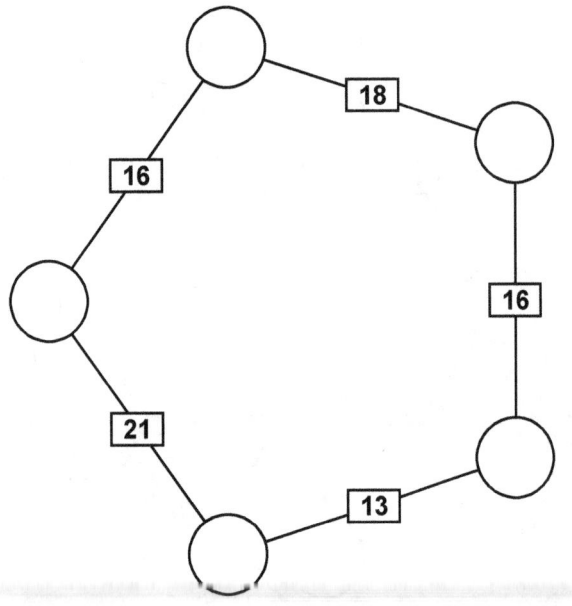

Puzzle 72

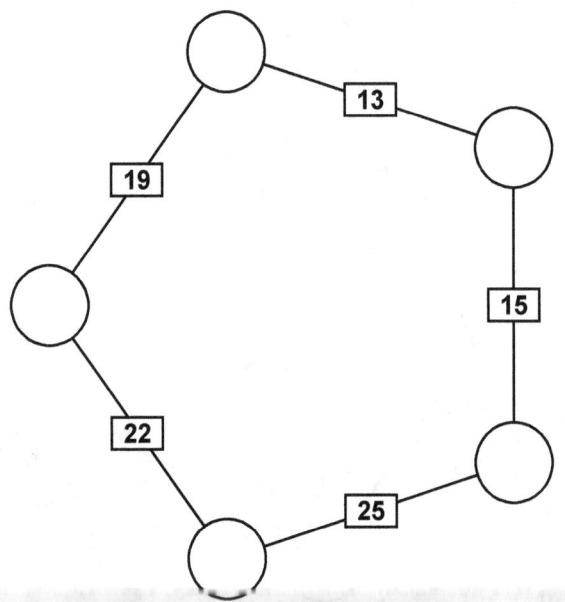

Puzzle 73

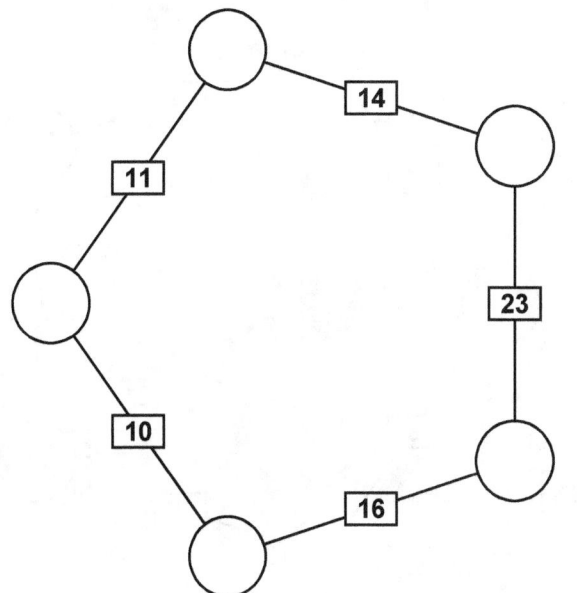

Puzzle 74

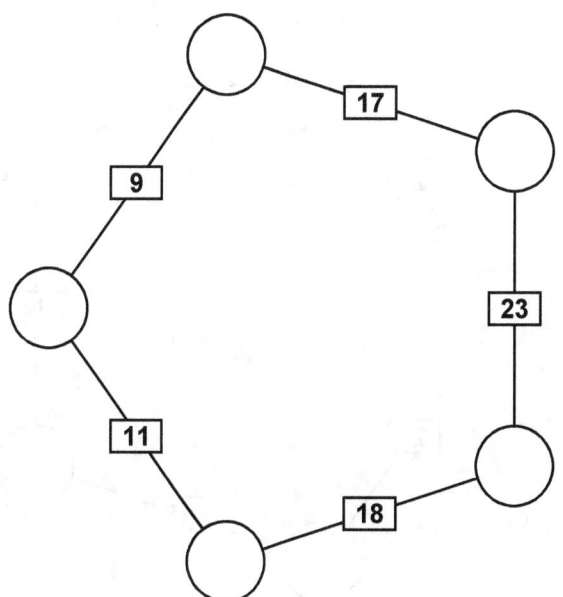

Puzzle 75

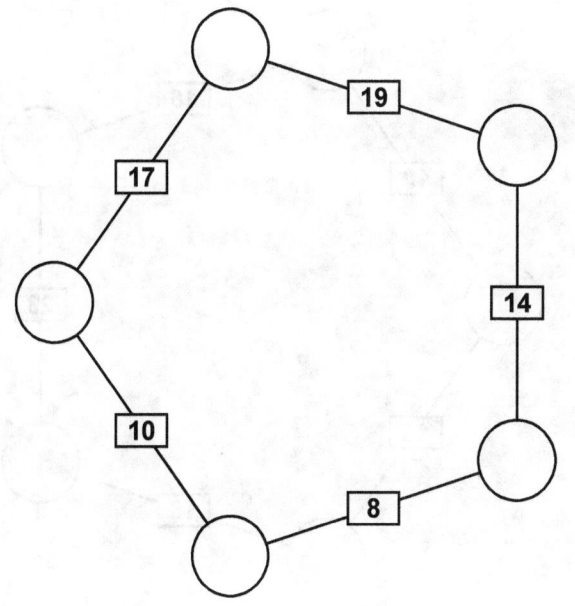

Puzzle 76

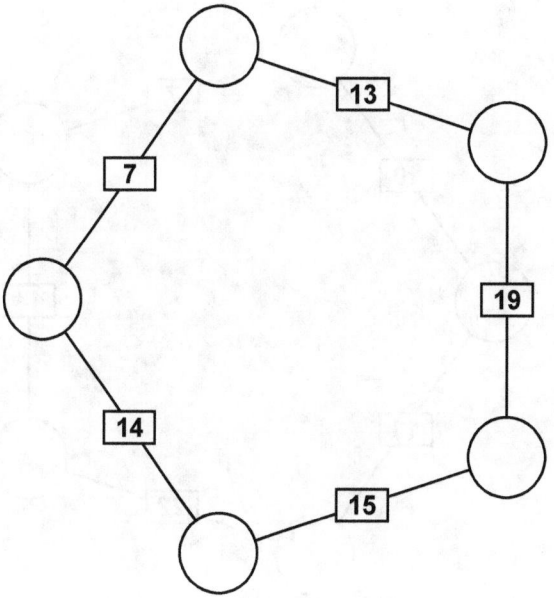

Puzzle 77

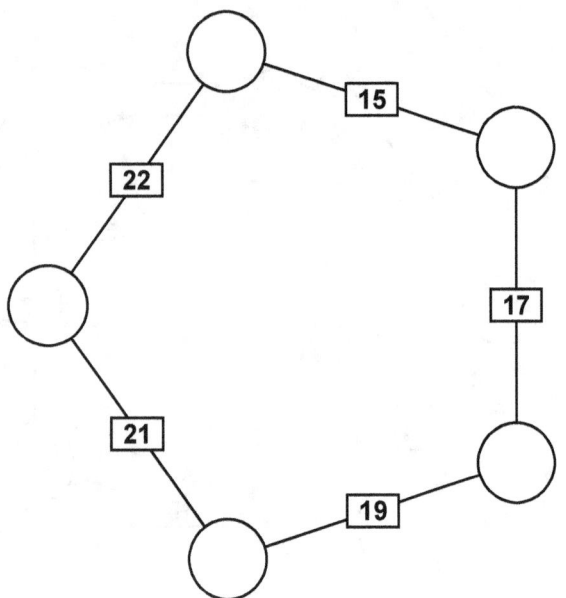

Puzzle 78

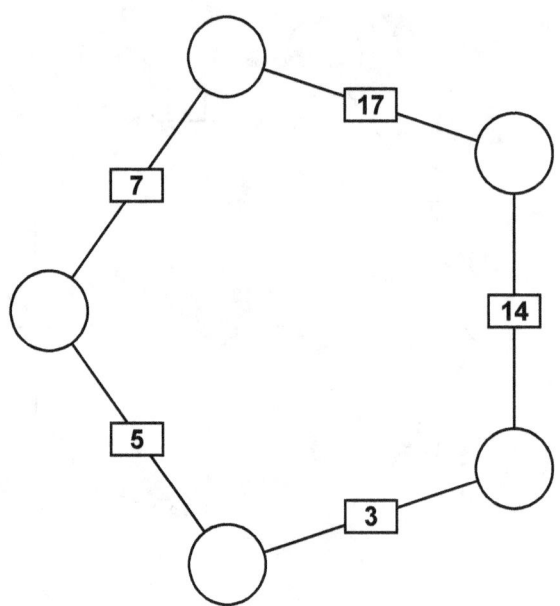

Puzzle 79

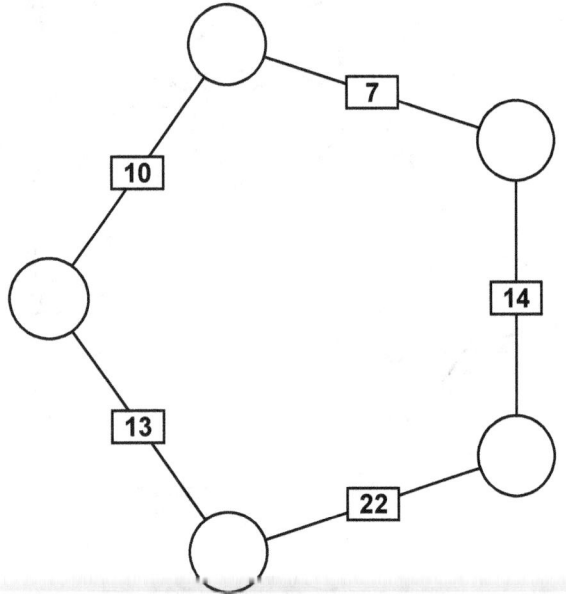

Puzzle 80

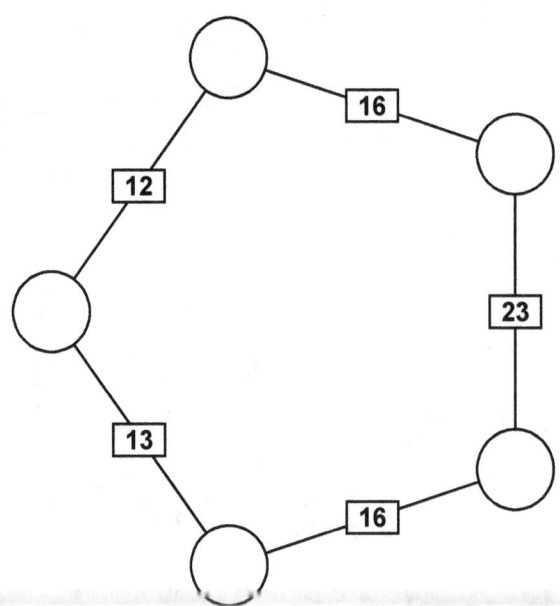

Puzzle 81

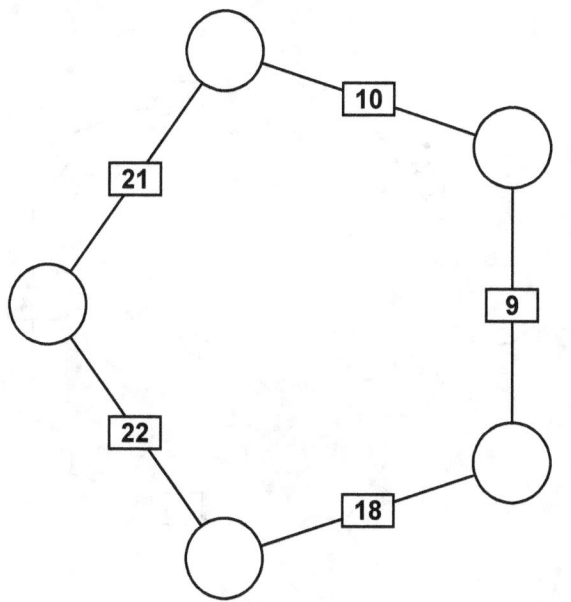

Puzzle 82

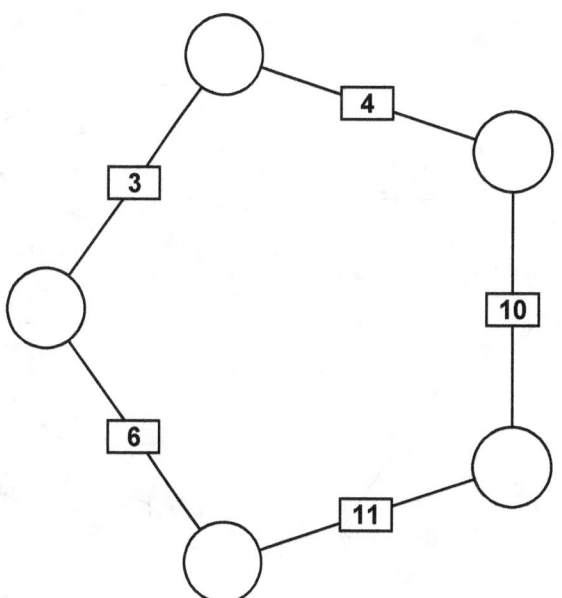

Puzzle 83

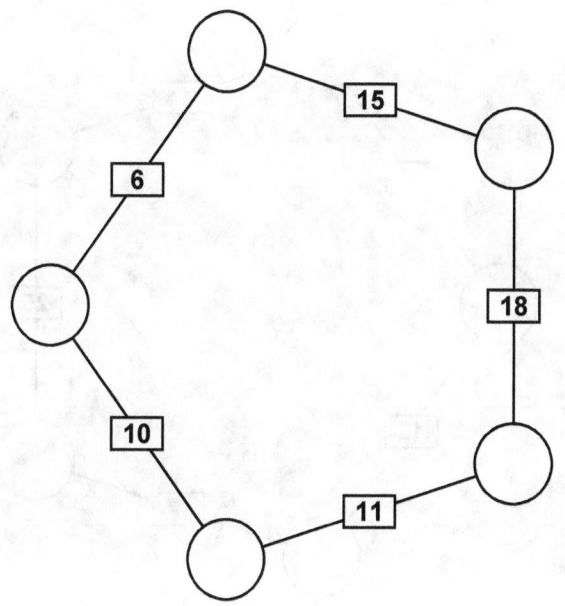

Puzzle 84

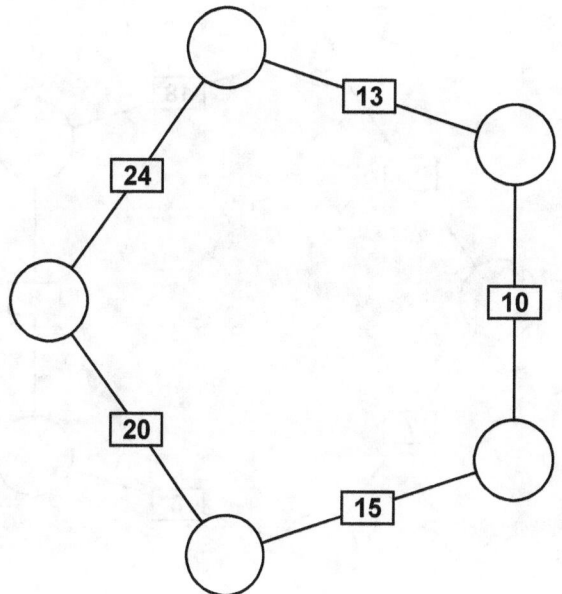

Puzzle 85

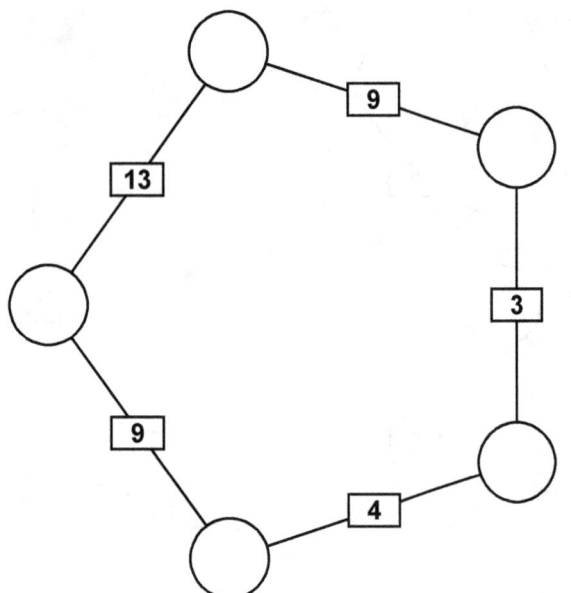

Puzzle 86

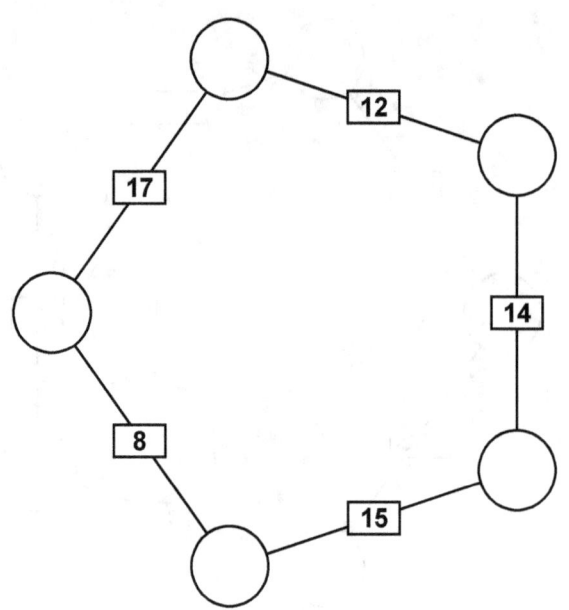

Puzzle 87

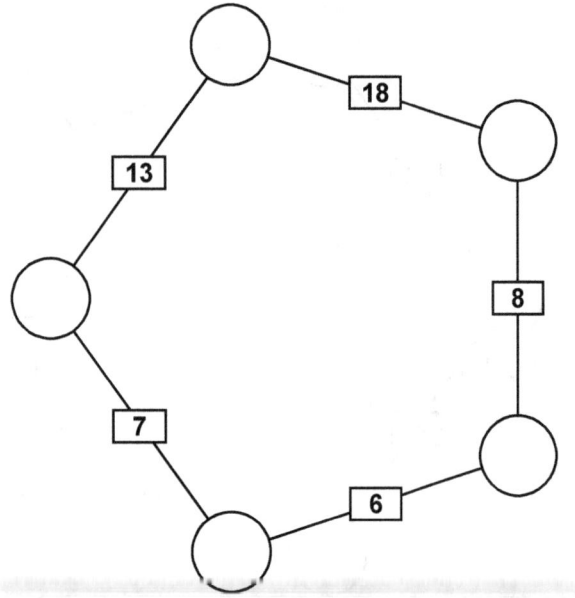

Puzzle 88

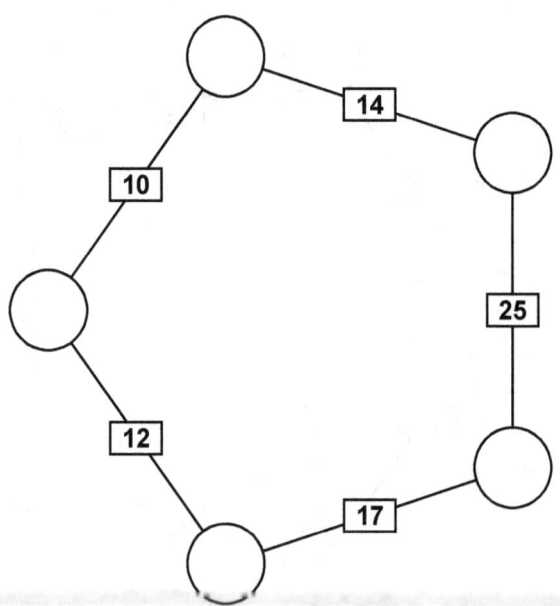

Puzzle 89

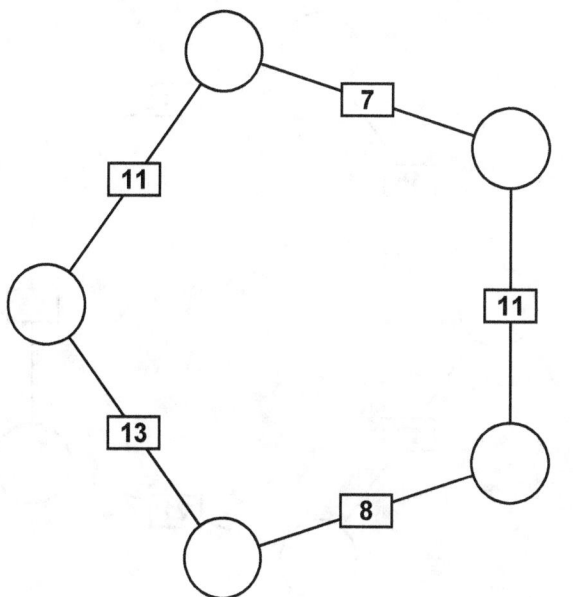

Puzzle 90

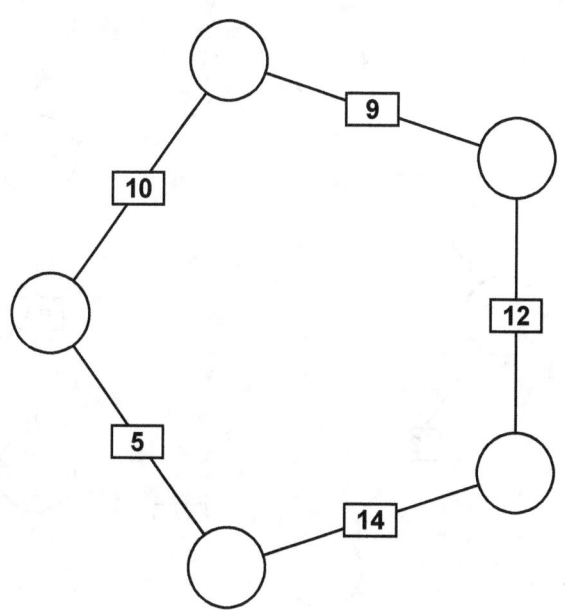

Puzzle 91

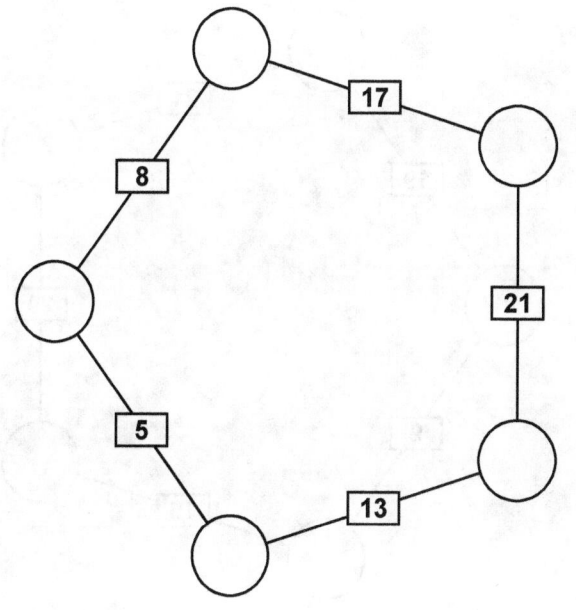

Puzzle 92

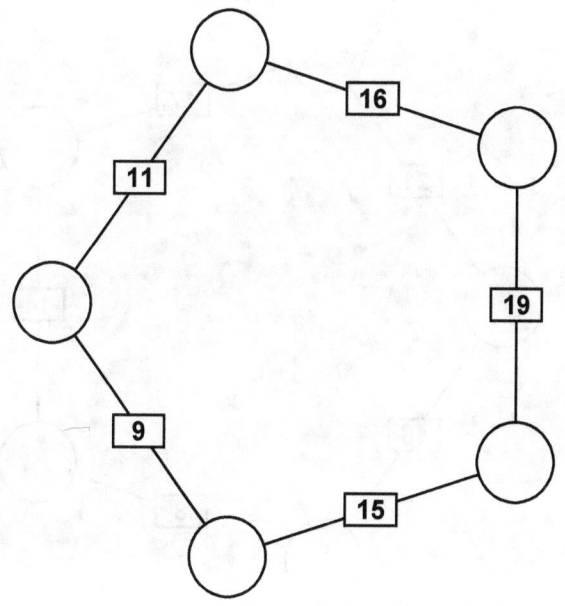

Puzzle 93

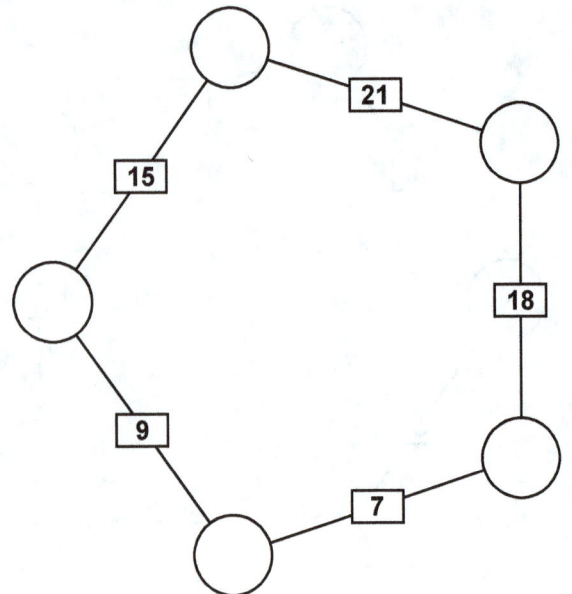

Puzzle 94

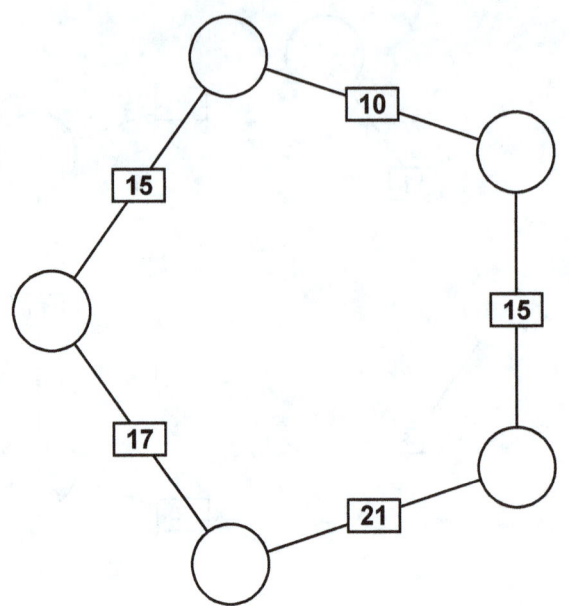

Puzzle 95

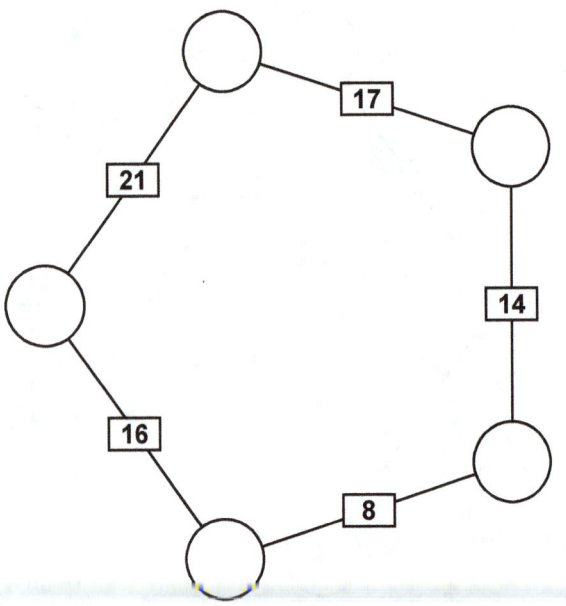

Puzzle 96

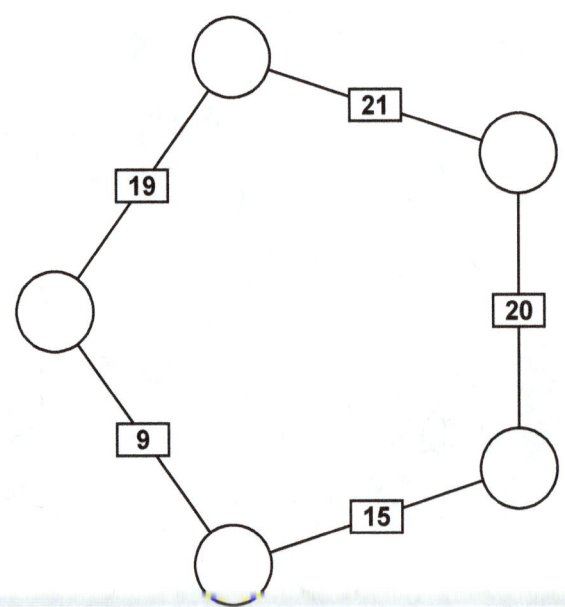

Puzzle 97

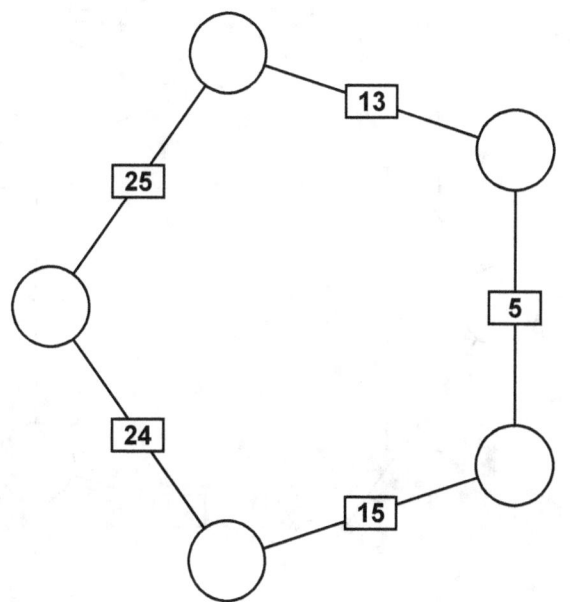

Puzzle 98

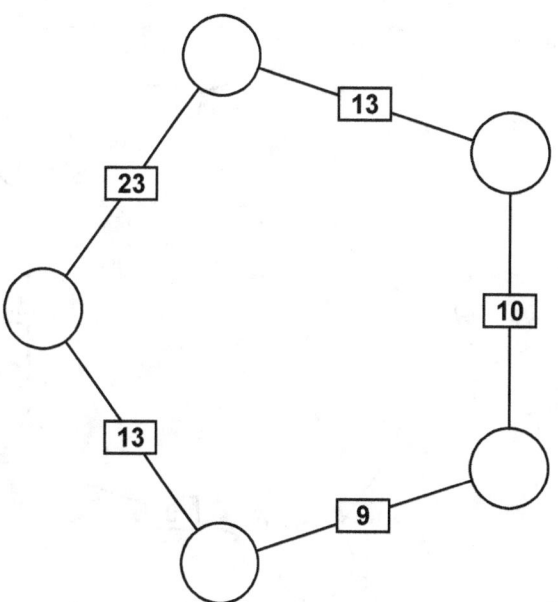

Puzzle 99

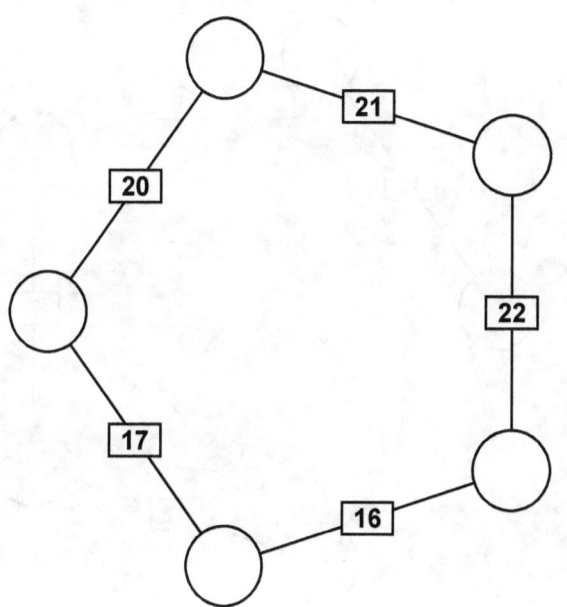

Puzzle 100

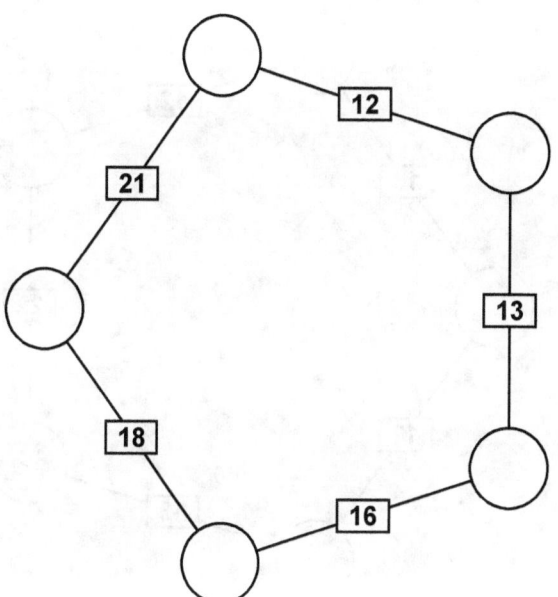

Puzzle 101

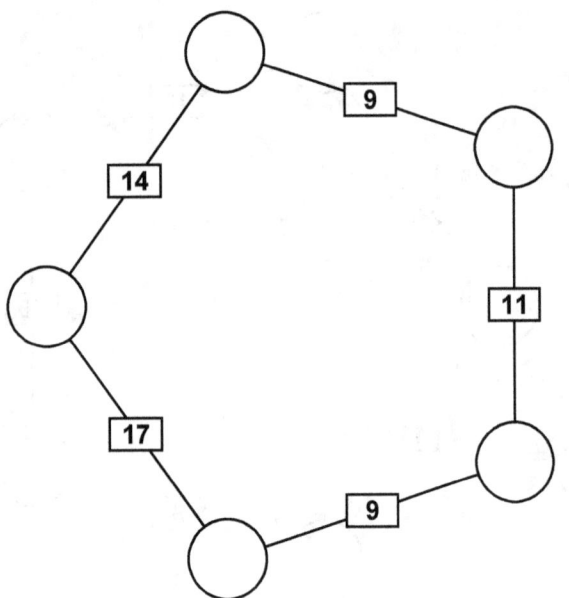

Puzzle 102

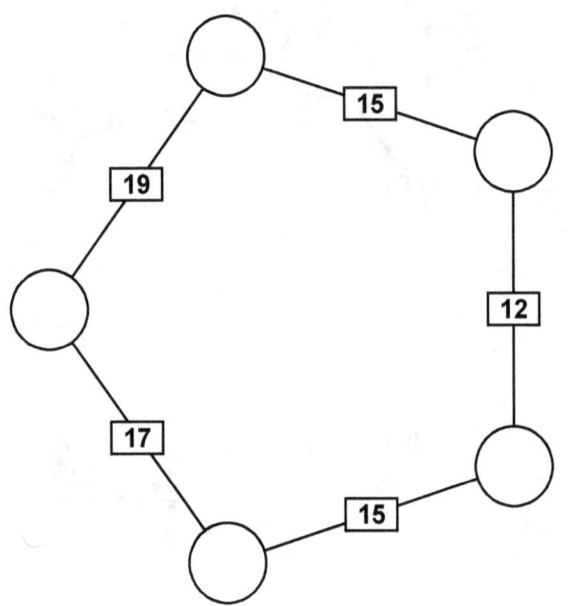

Puzzle 103

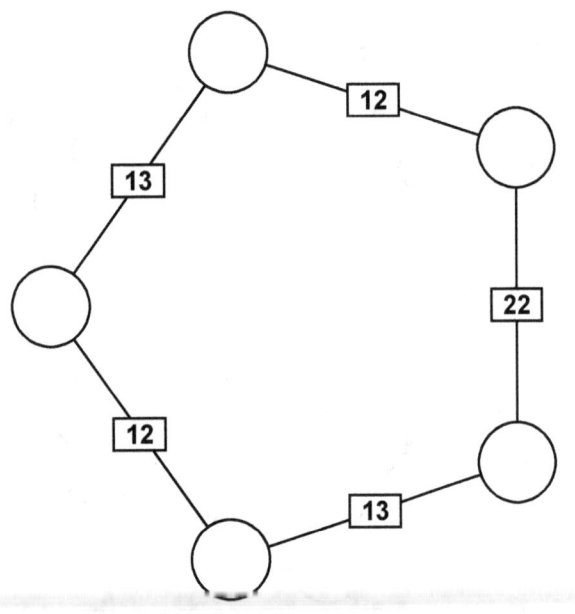

Puzzle 104

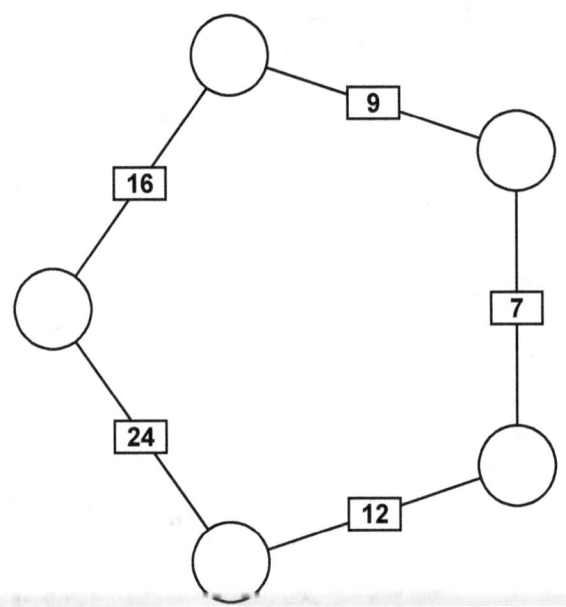

Puzzle 105

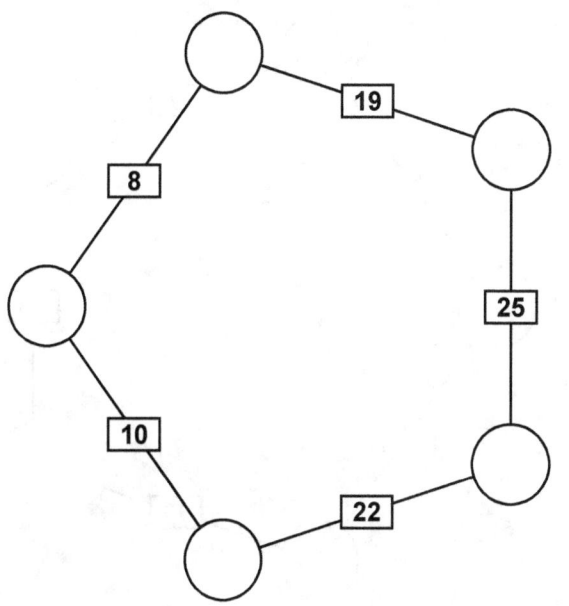

Puzzle 106

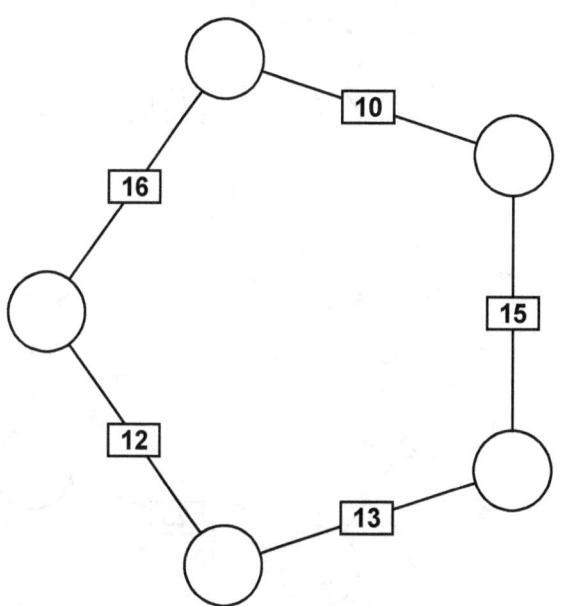

Puzzle 107

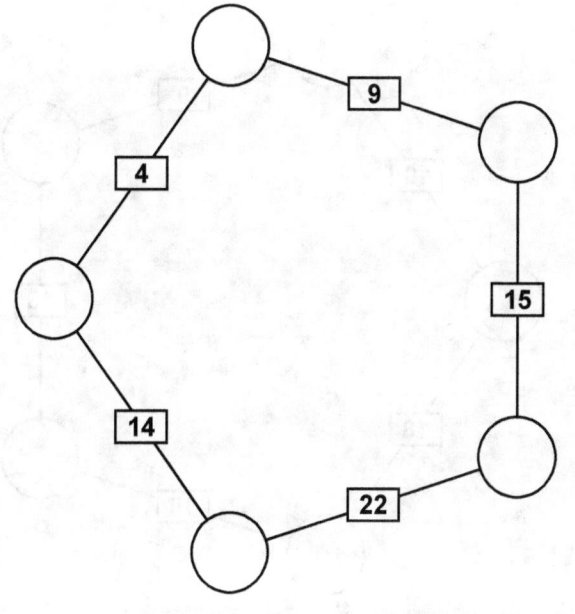

Puzzle 108

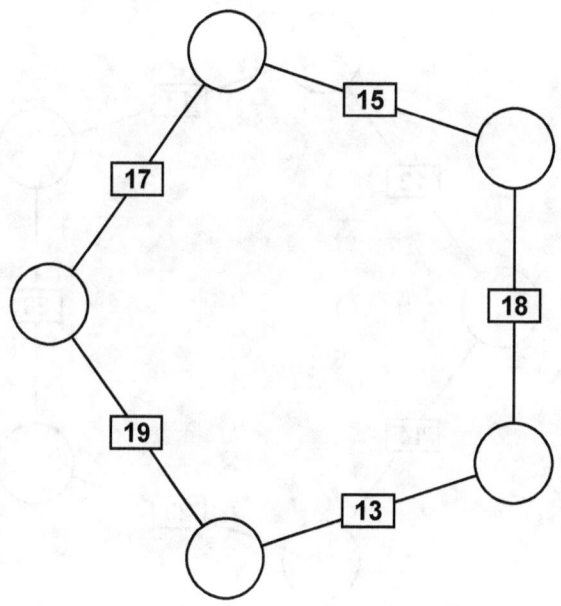

Puzzle 109

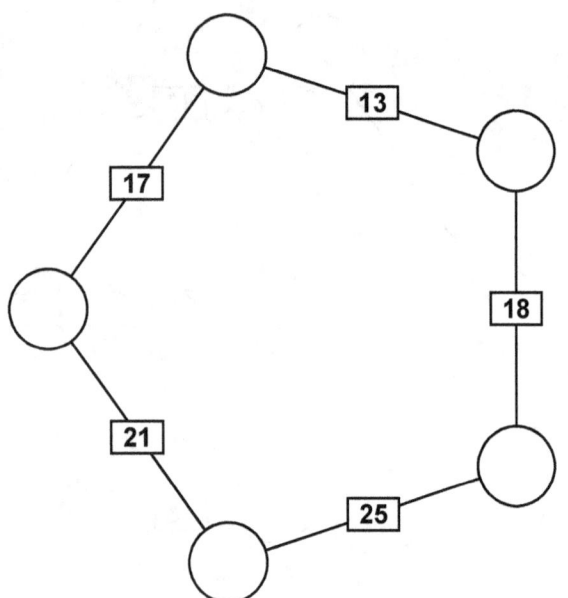

Puzzle 110

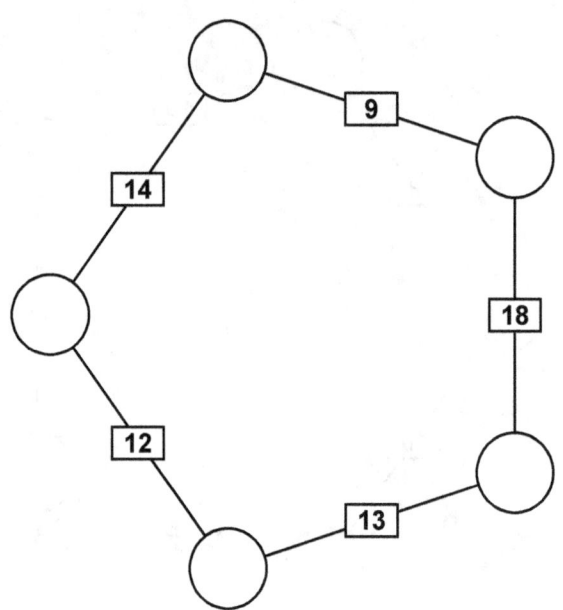

Puzzle 111

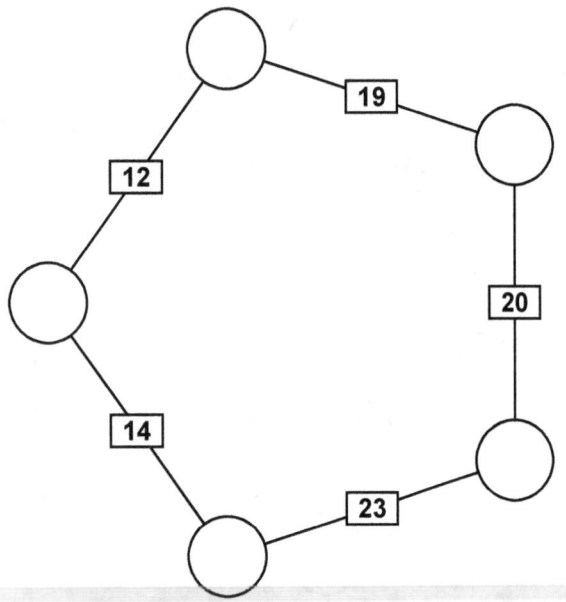

Puzzle 112

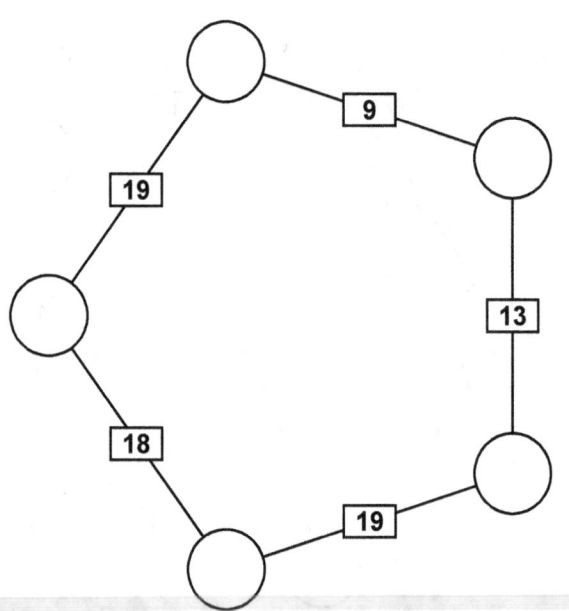

Puzzle 113

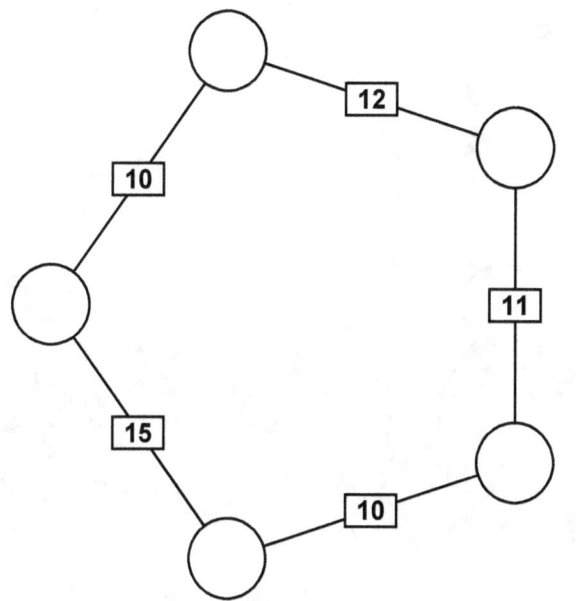

Puzzle 114

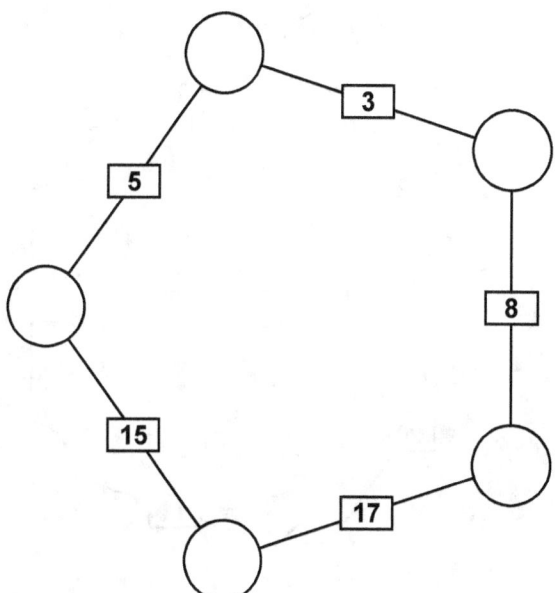

Puzzle 115

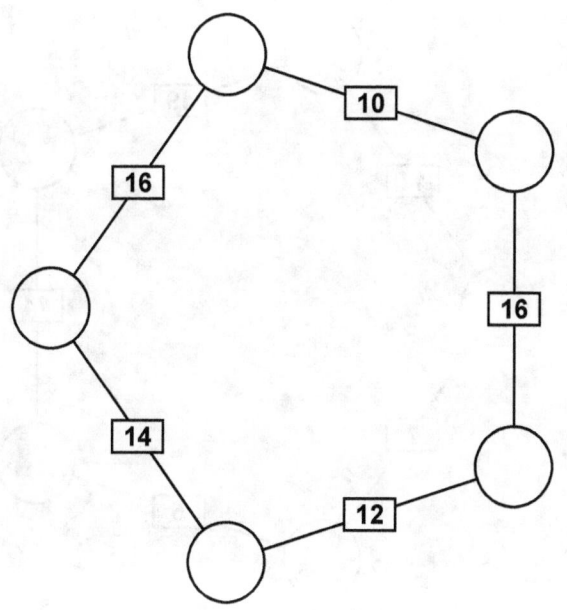

Puzzle 116

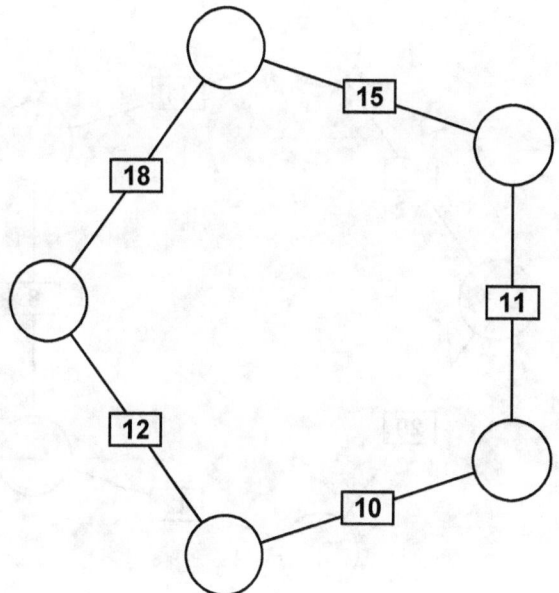

Puzzle 117

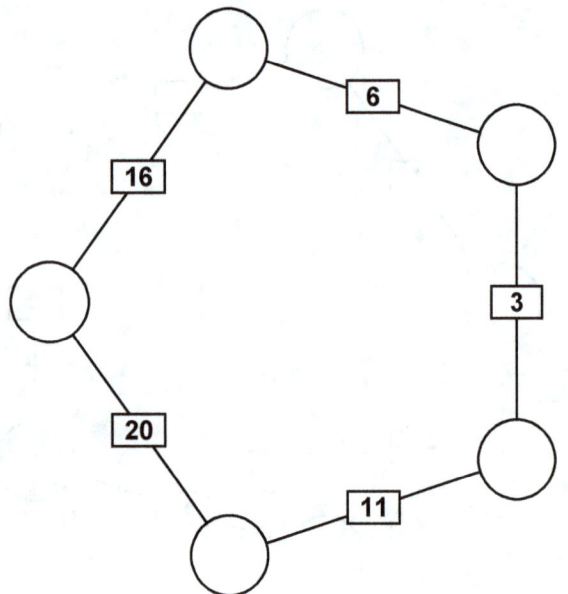

Puzzle 118

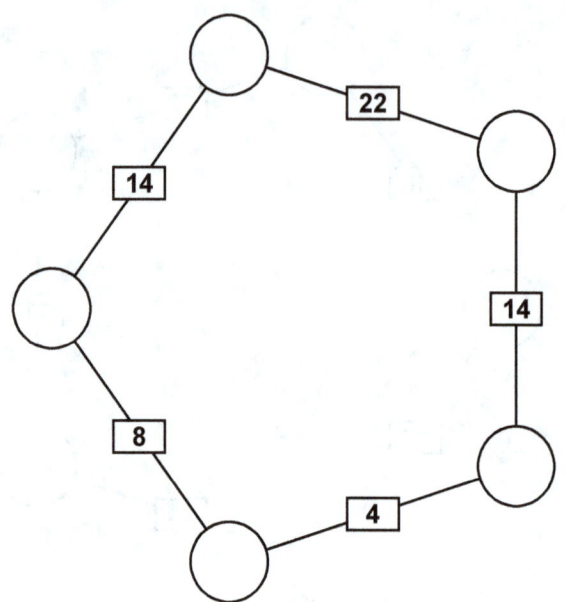

Puzzle 119

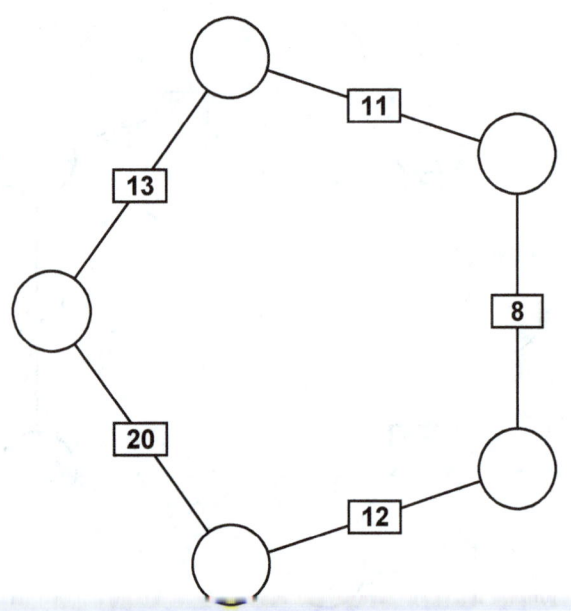

Puzzle 120

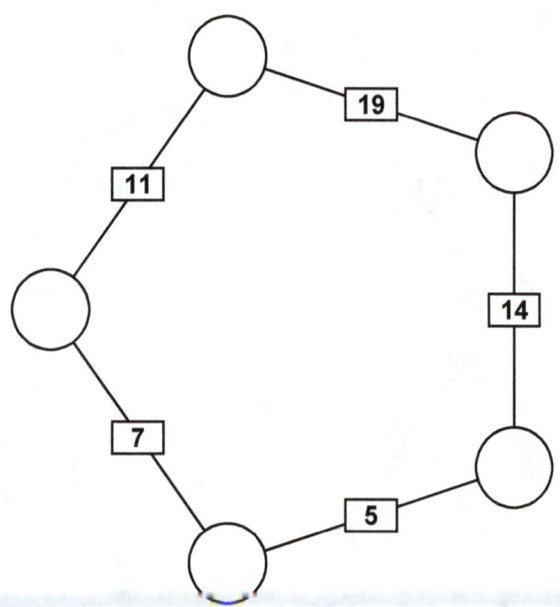

Puzzle 121

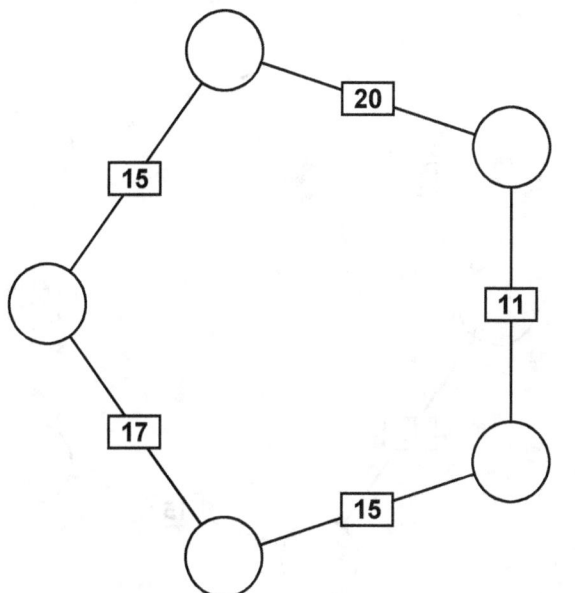

Puzzle 122

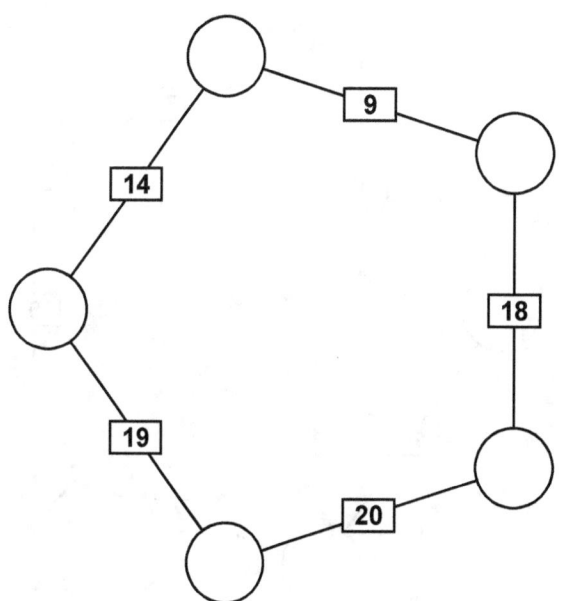

Puzzle 123

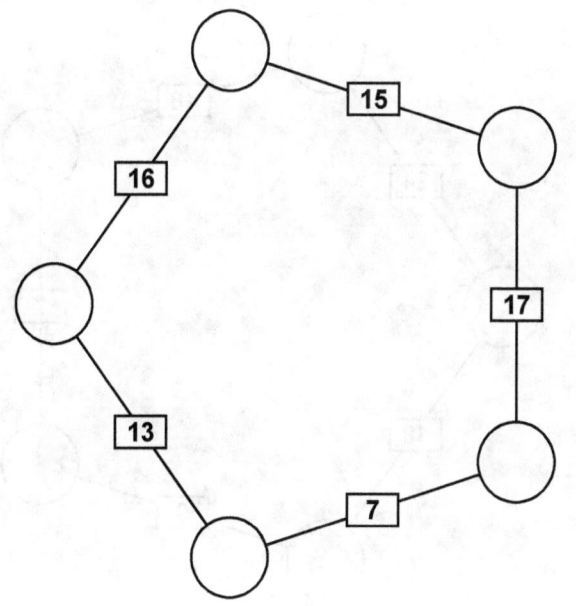

Puzzle 124

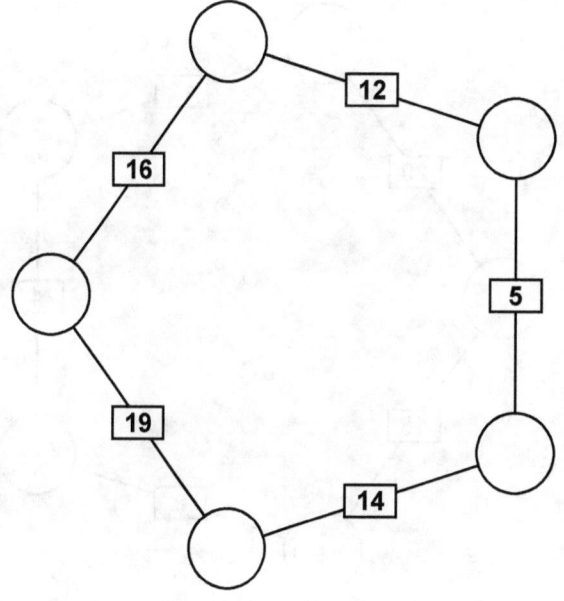

Puzzle 125

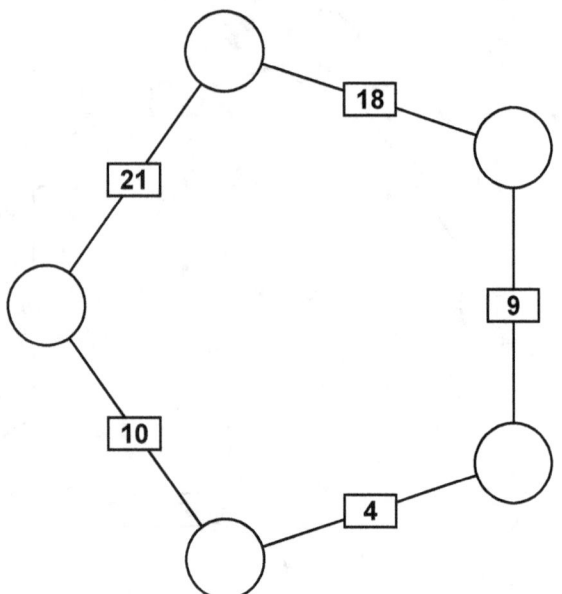

Puzzle 126

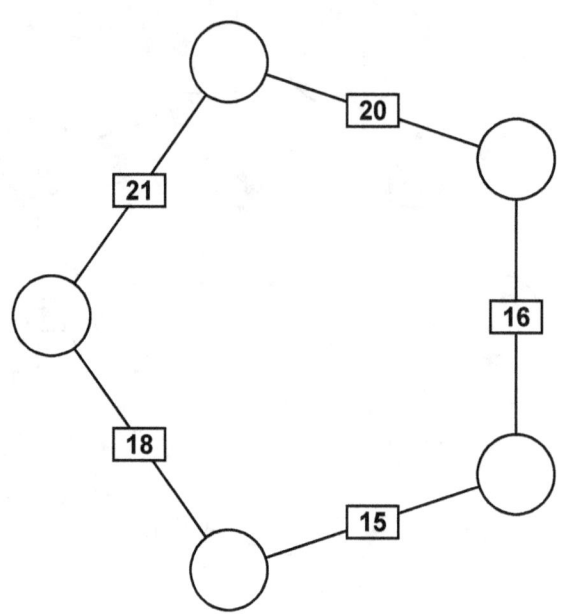

Puzzle 127

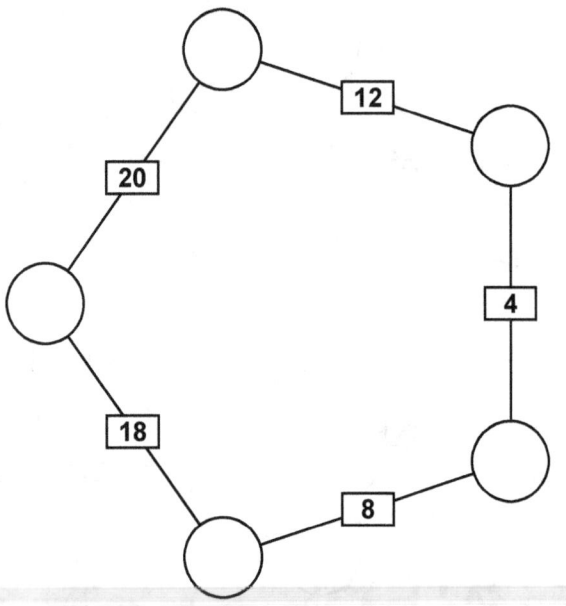

Puzzle 128

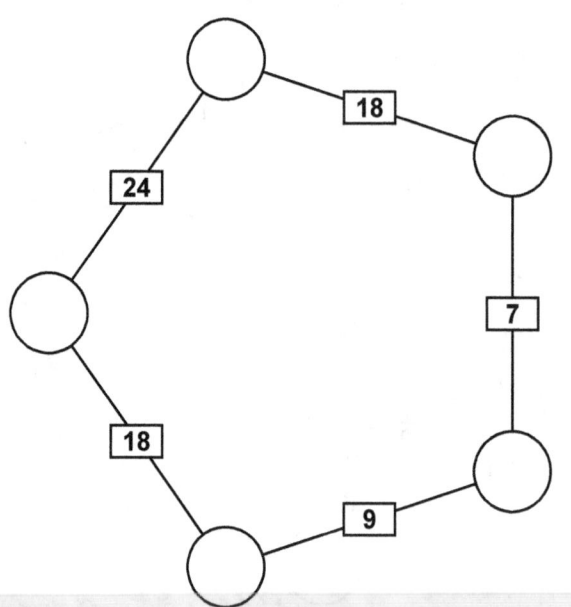

Puzzle 129

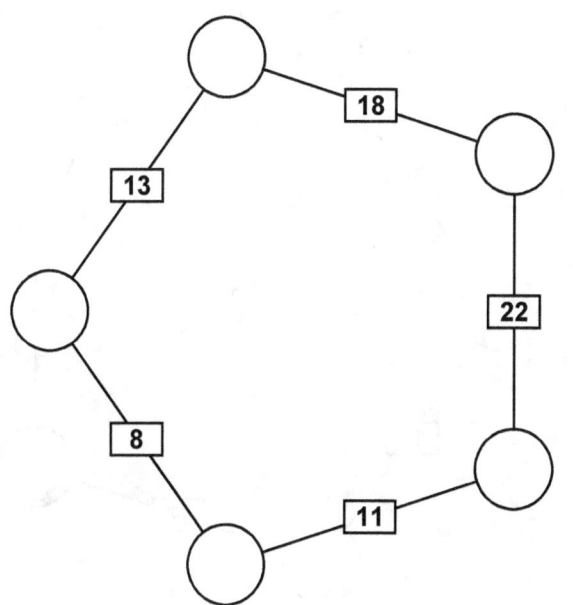

Puzzle 130

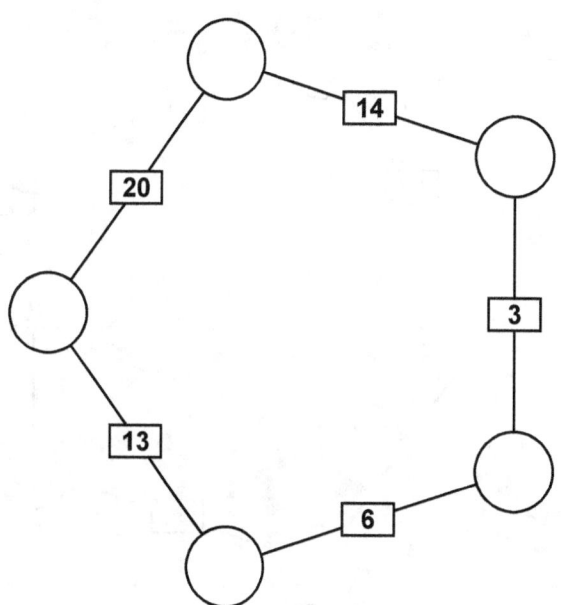

Puzzle 131

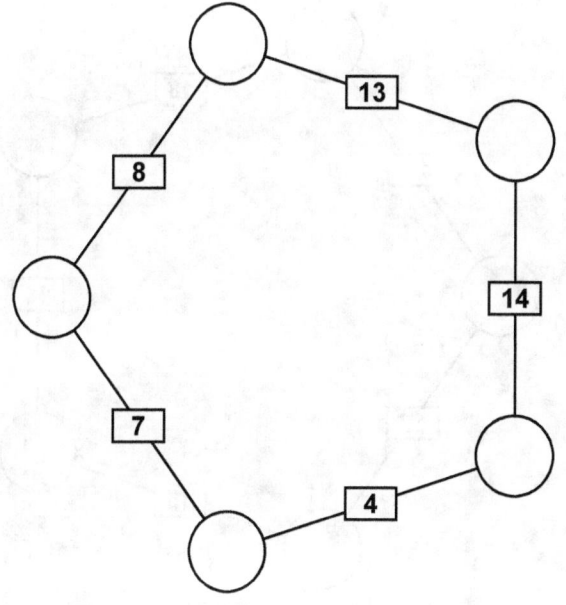

Puzzle 132

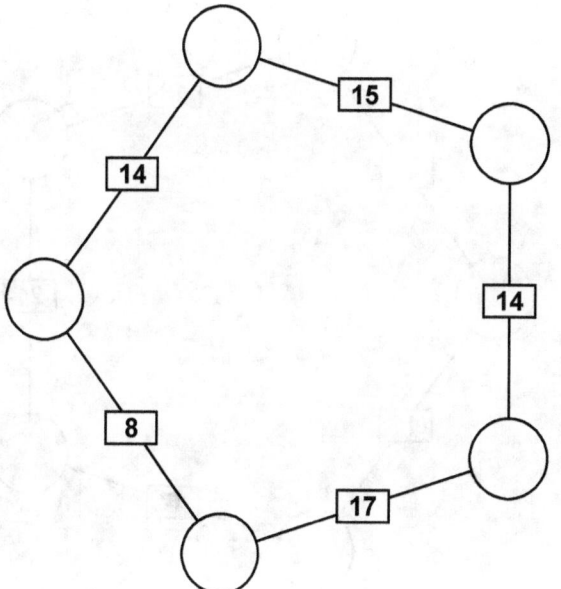

Puzzle 133

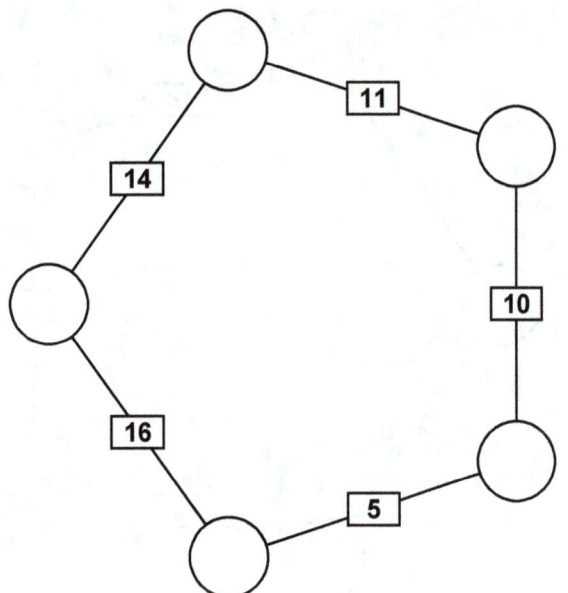

Puzzle 134

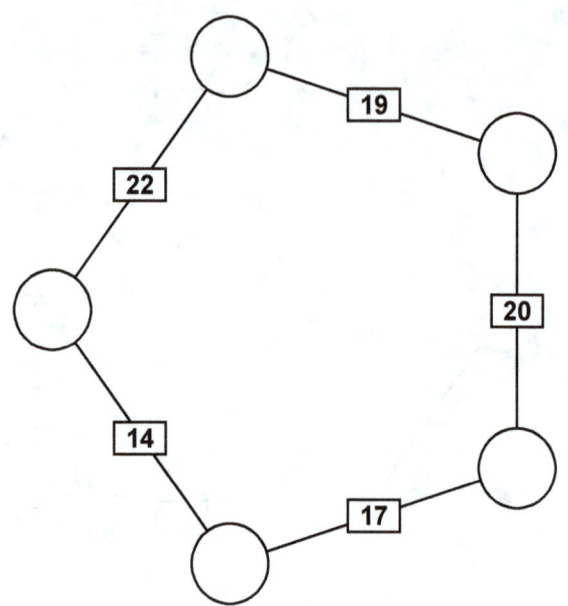

Puzzle 135

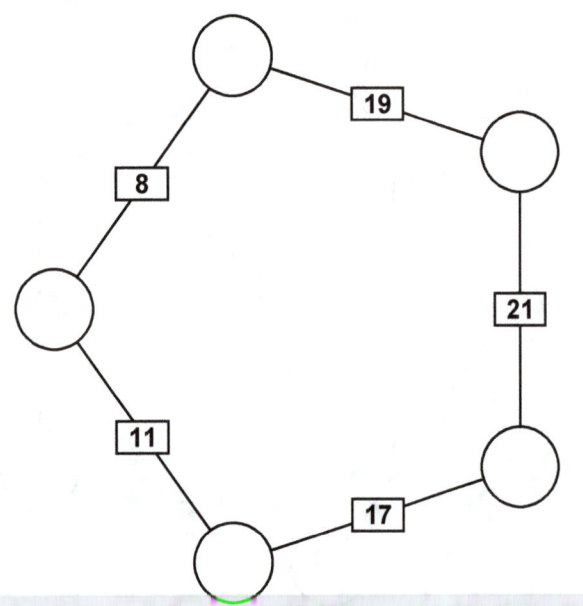

Puzzle 136

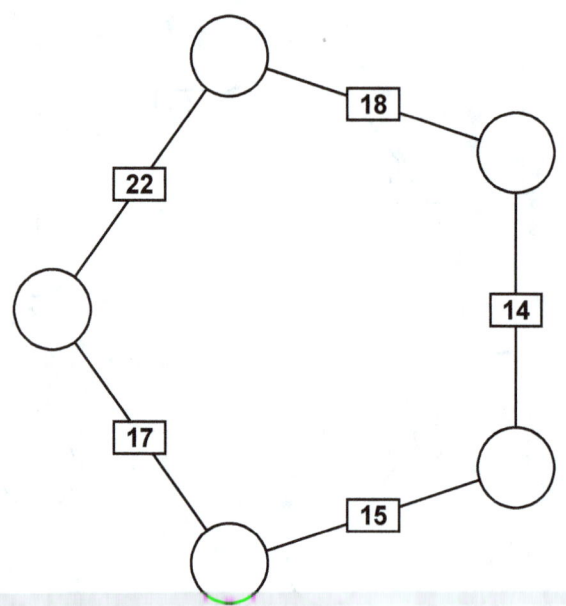

Puzzle 137

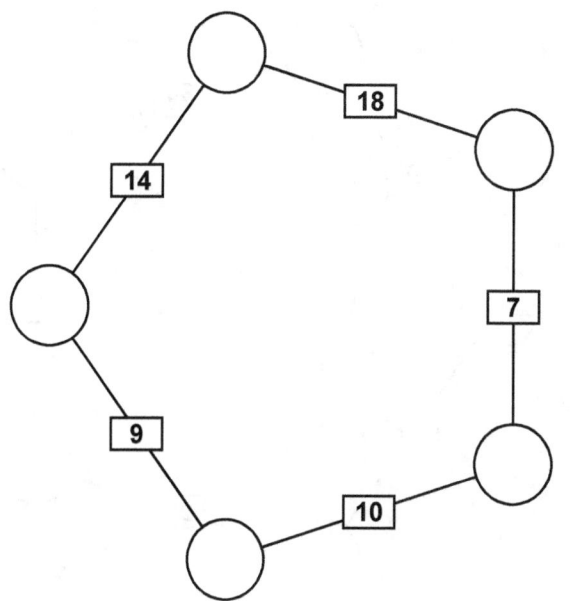

Puzzle 138

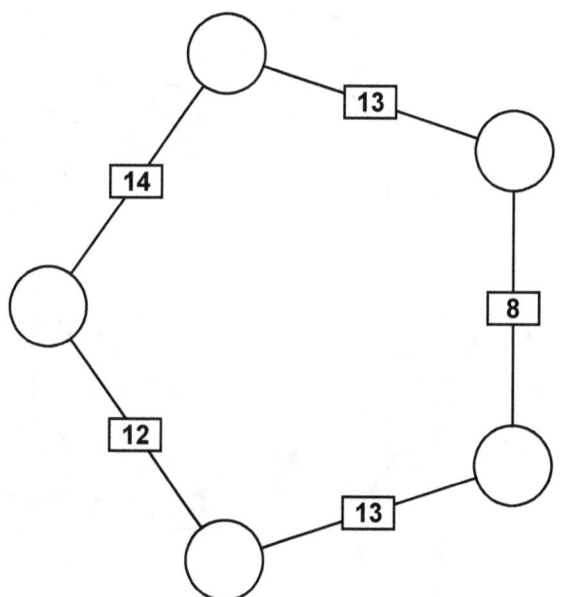

Puzzle 139

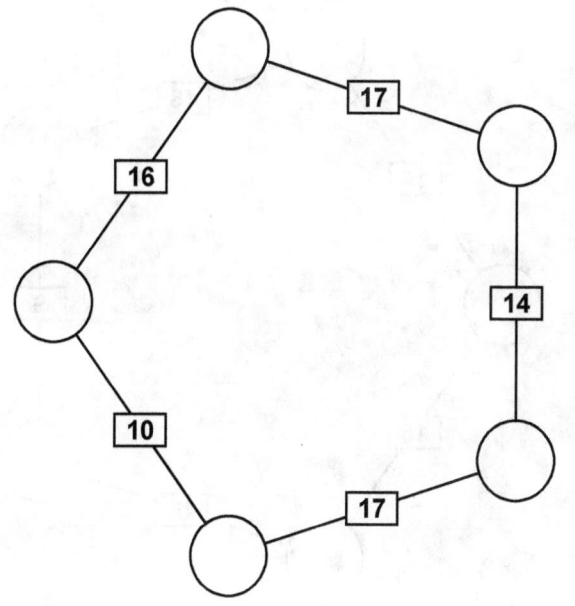

Puzzle 140

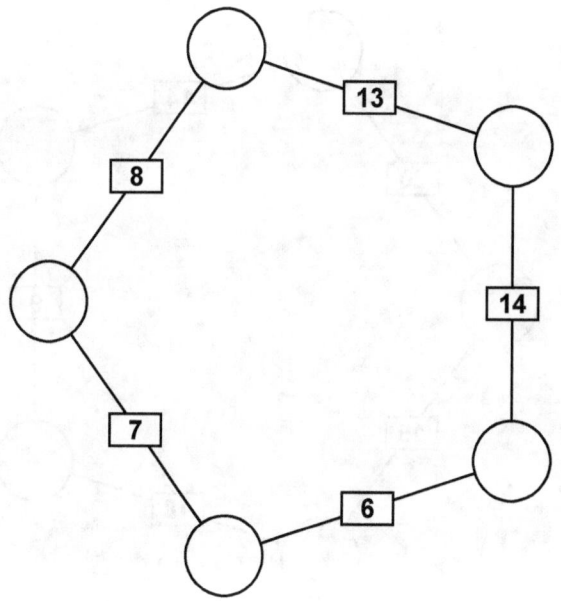

Puzzle 141

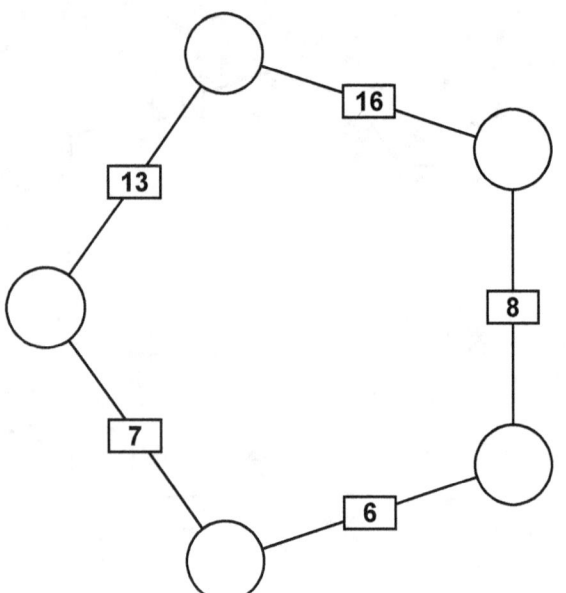

Puzzle 142

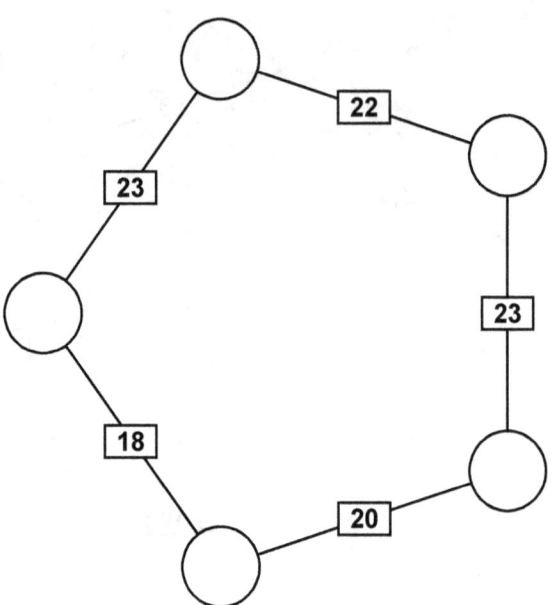

Puzzle 143

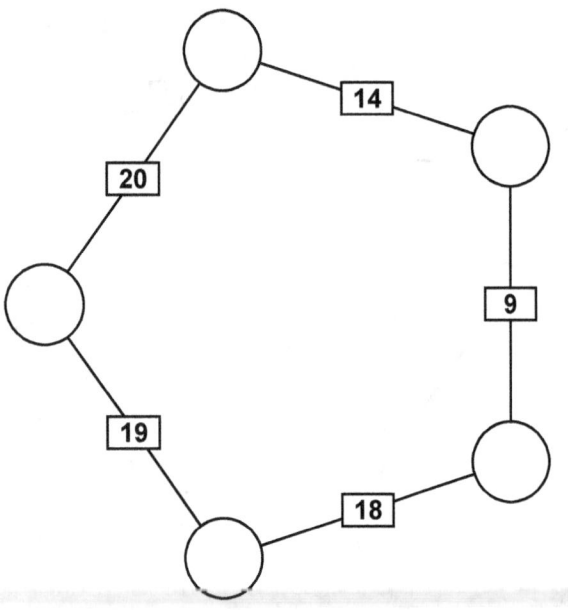

Puzzle 144

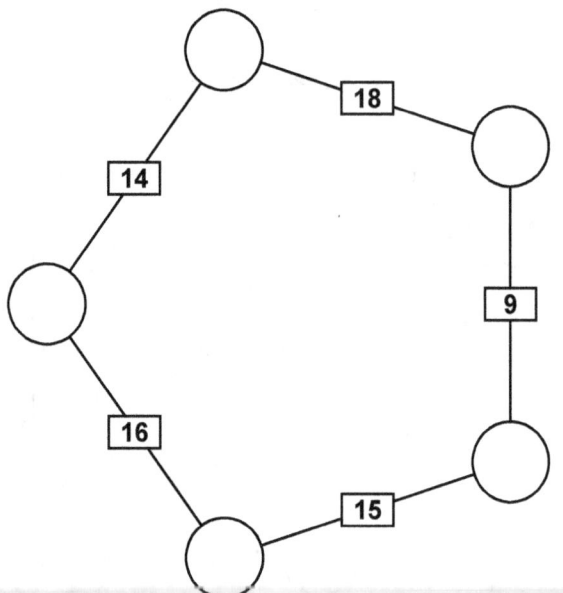

Puzzle 145

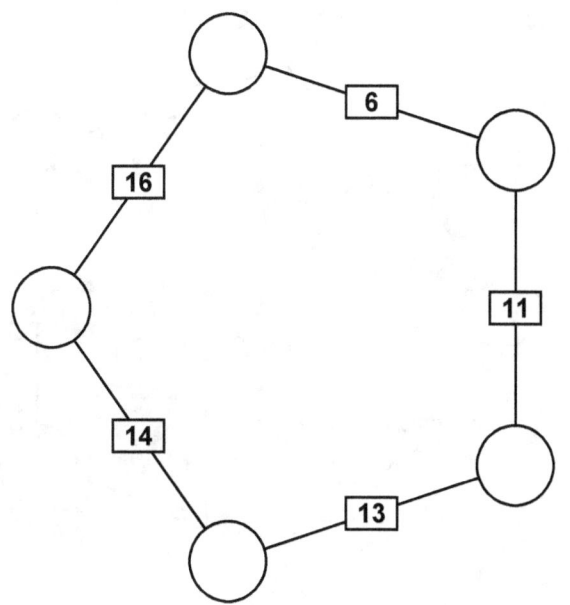

Puzzle 146

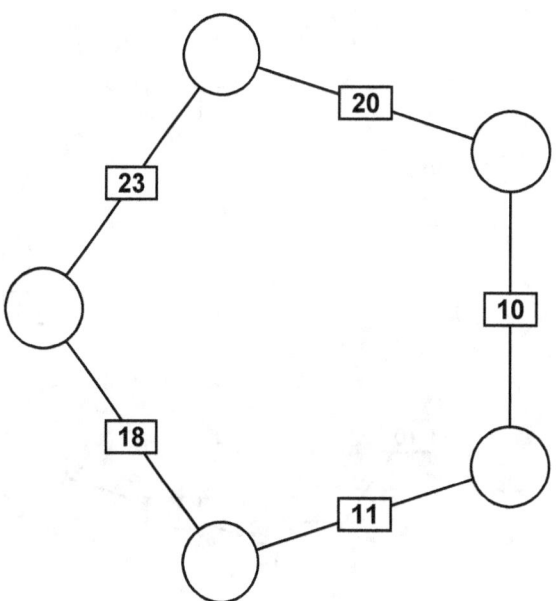

Puzzle 147

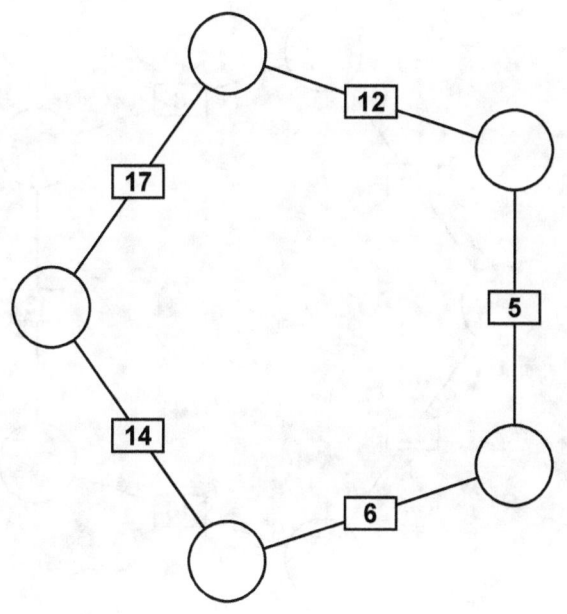

Puzzle 148

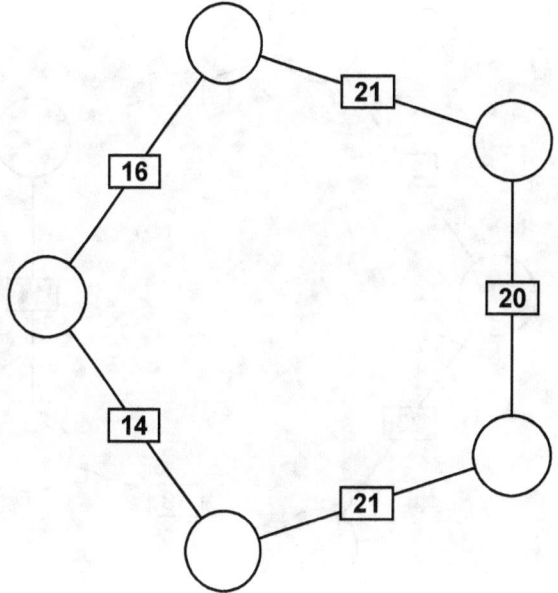

Puzzle 149

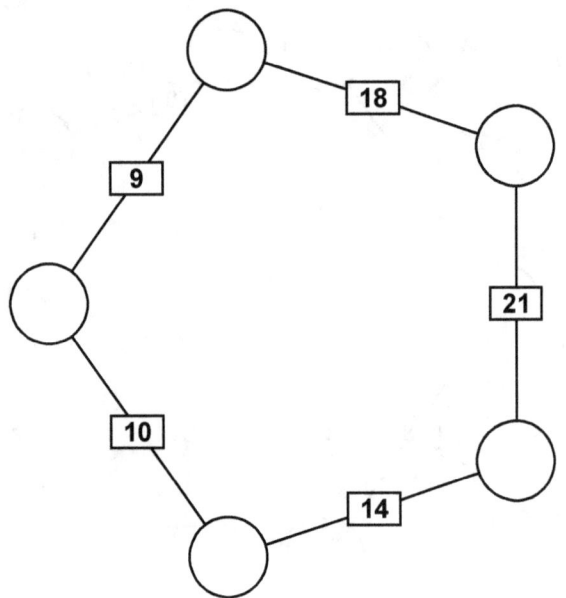

Puzzle 150

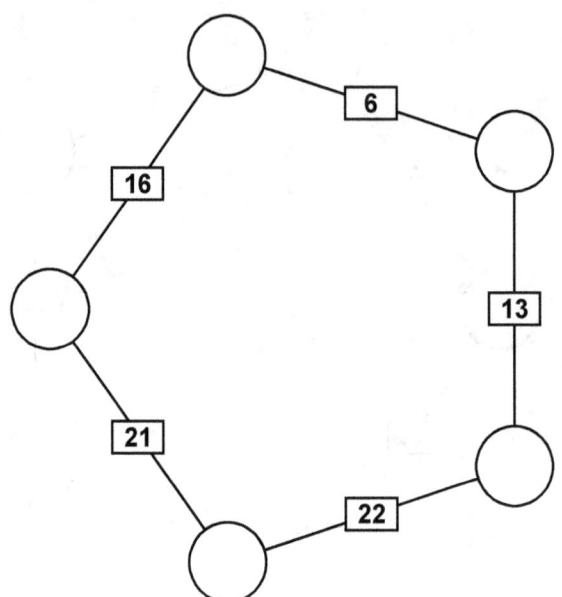

Puzzle 151

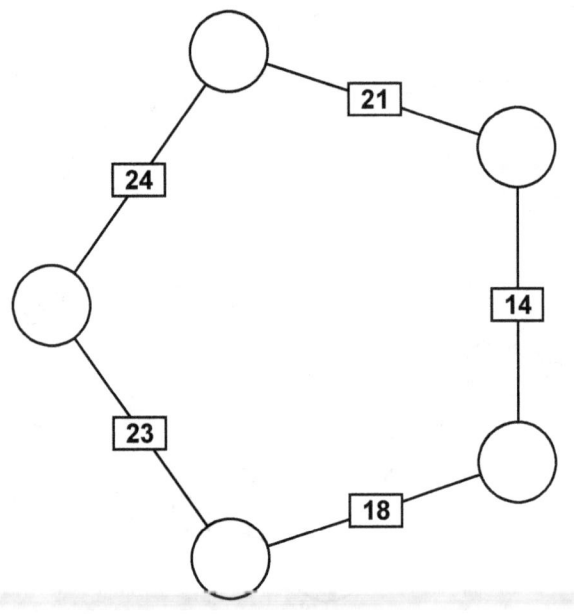

Puzzle 152

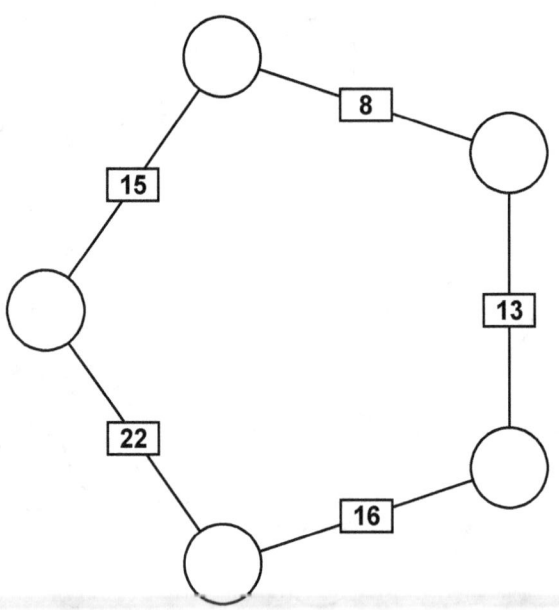

Puzzle 153

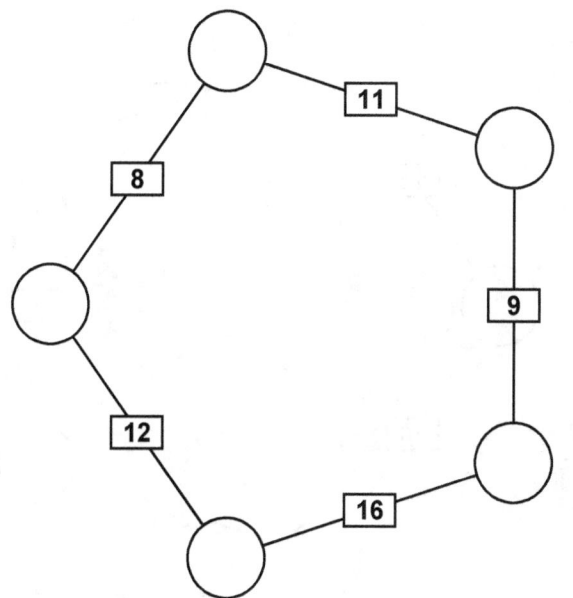

Puzzle 154

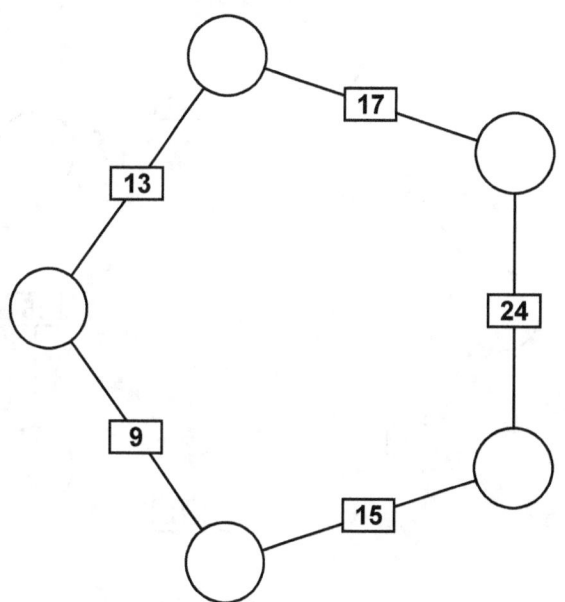

Puzzle 155

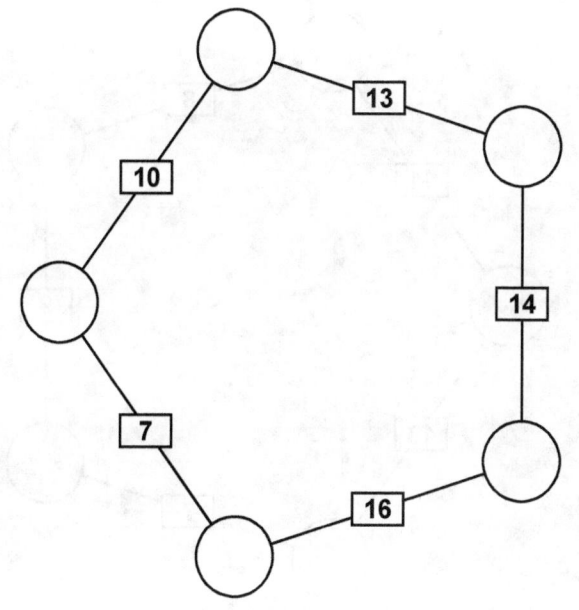

Puzzle 156

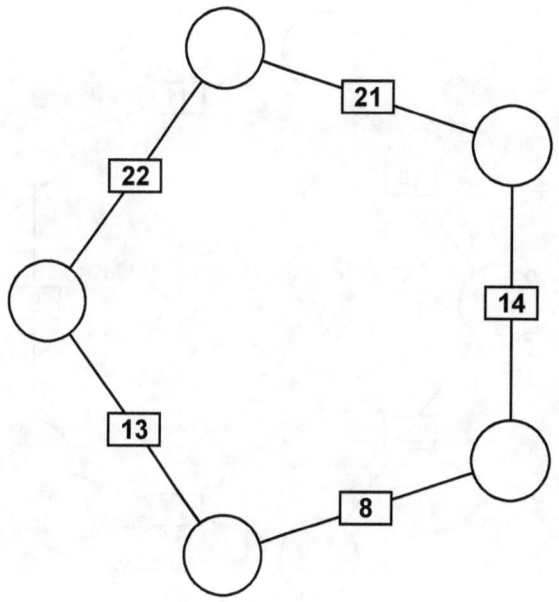

Puzzle 157

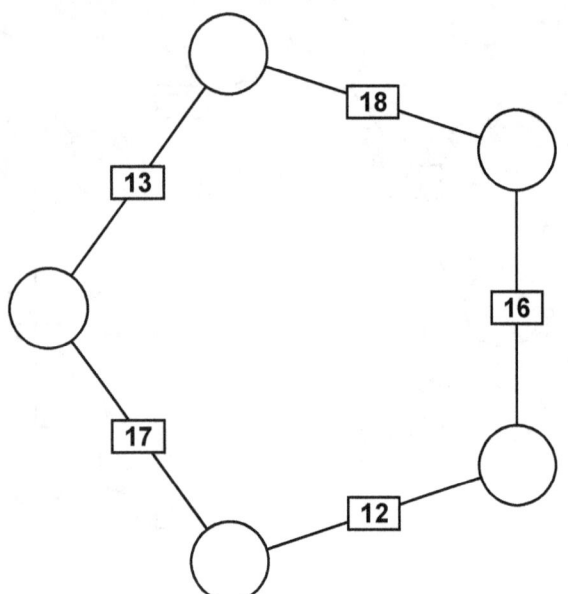

Puzzle 158

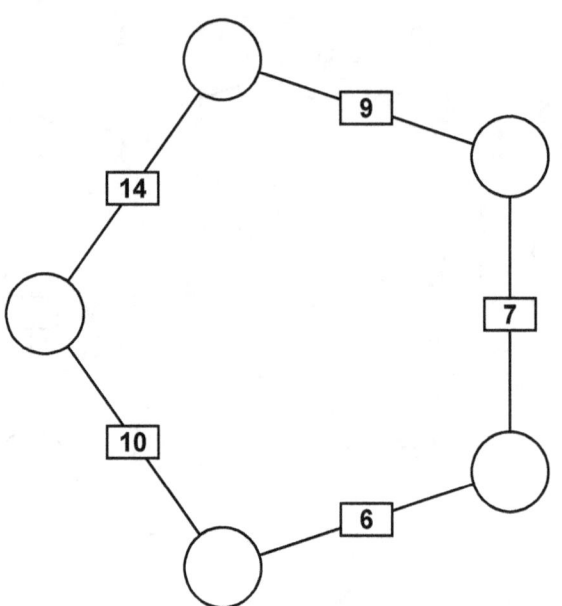

Puzzle 159

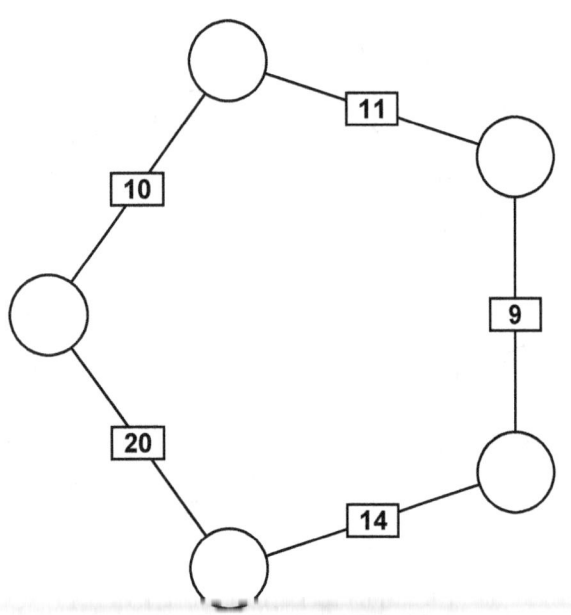

Puzzle 160

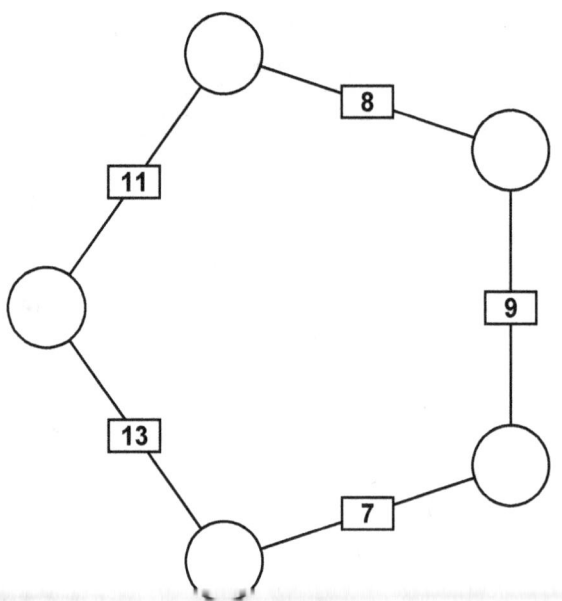

Puzzle 161

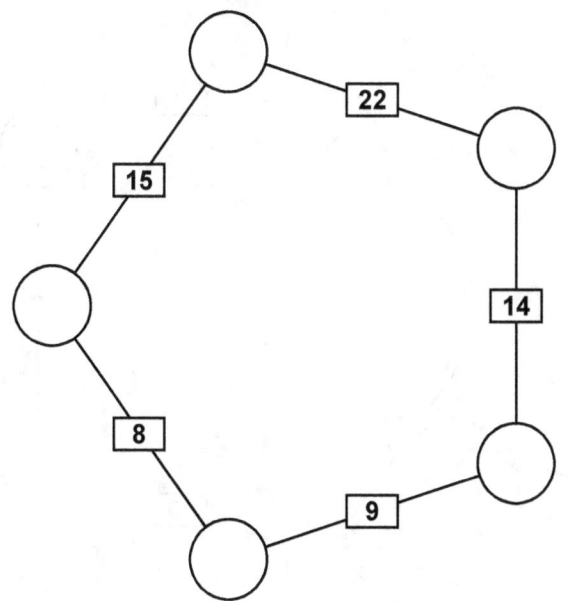

Puzzle 162

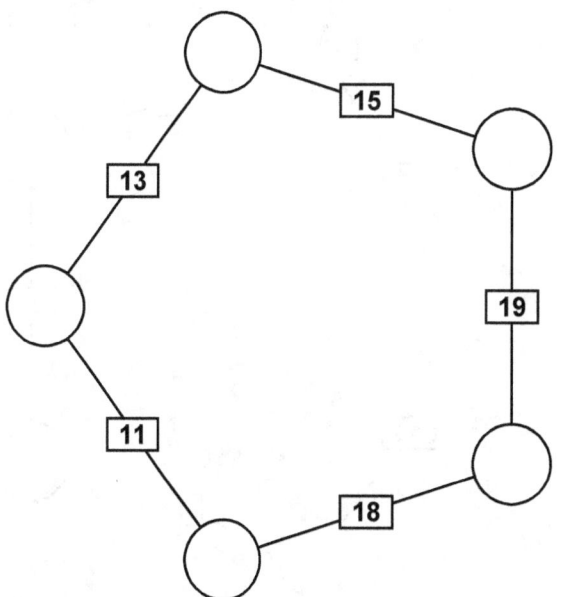

Puzzle 163

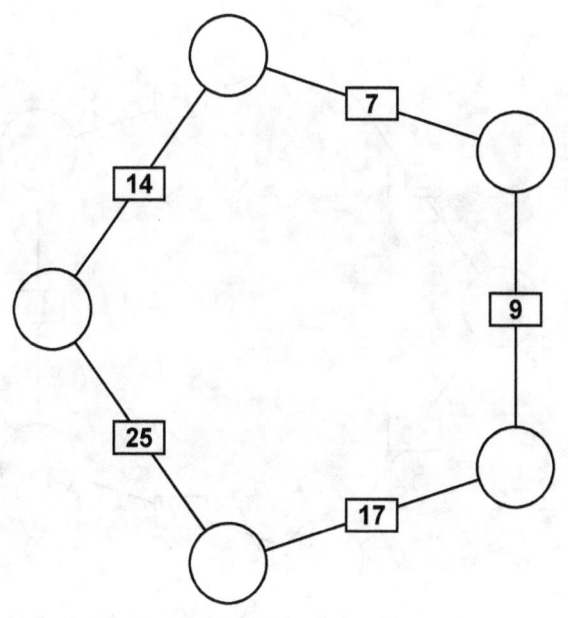

Puzzle 164

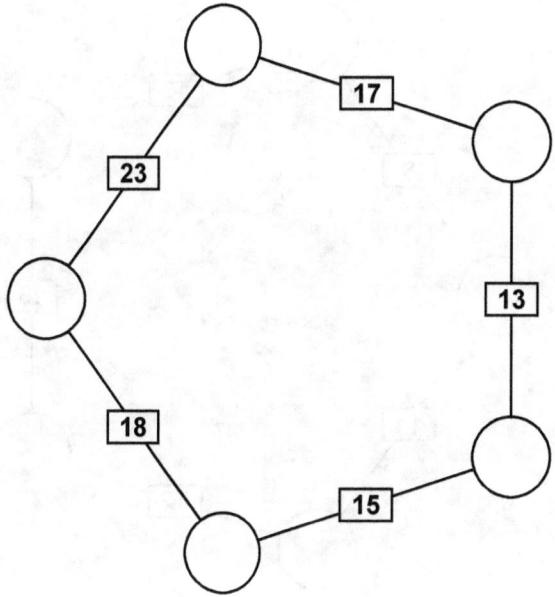

Puzzle 165

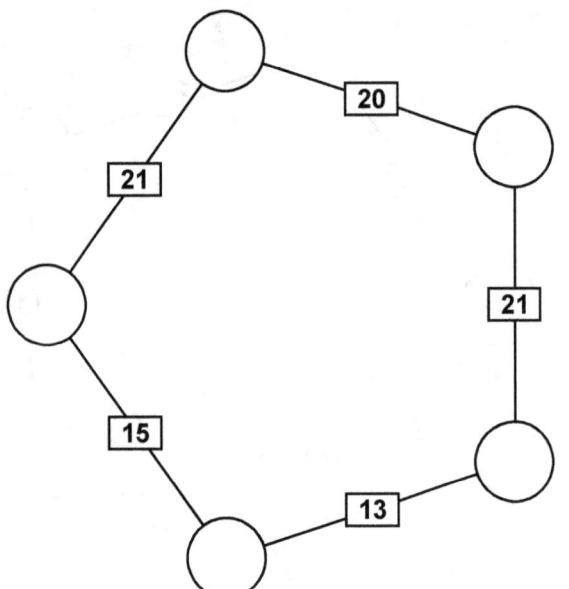

Puzzle 166

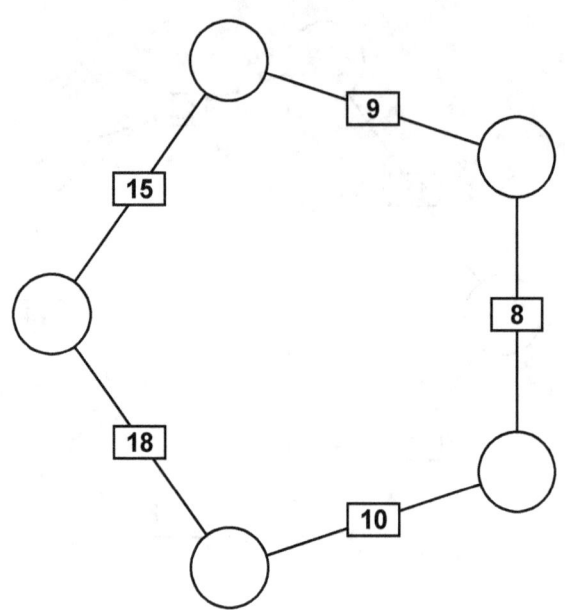

Puzzle 167

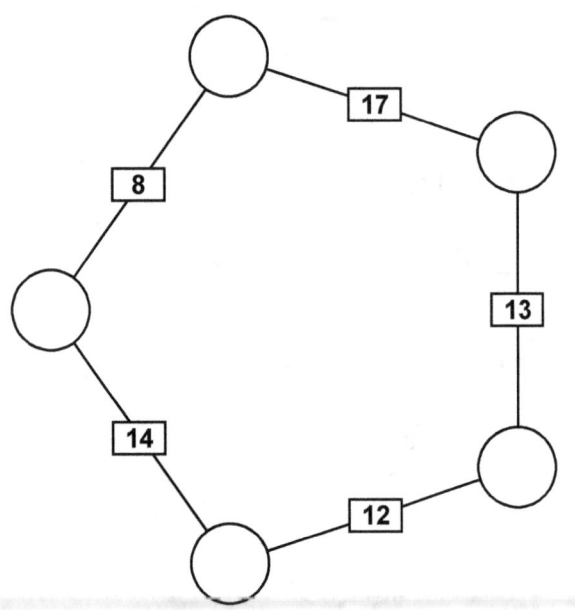

Puzzle 168

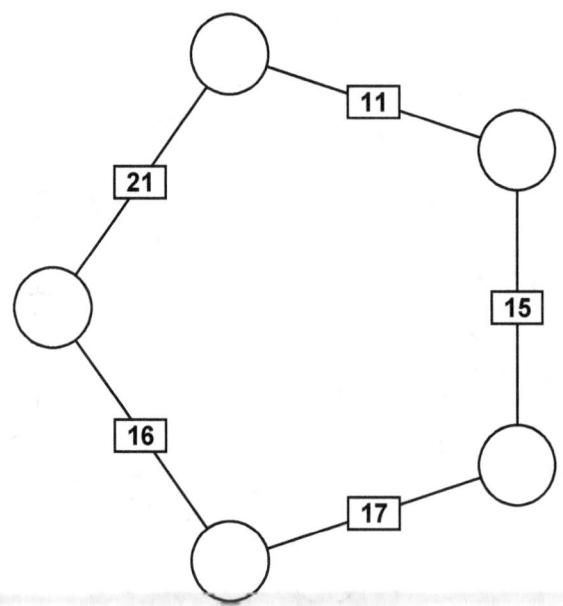

Puzzle 169

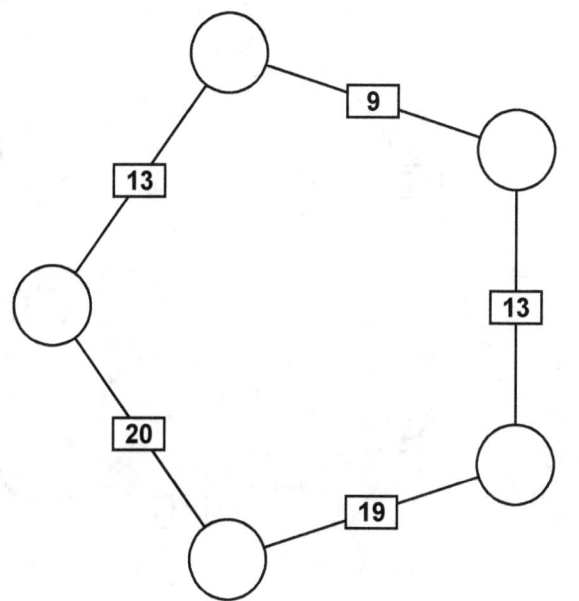

Puzzle 170

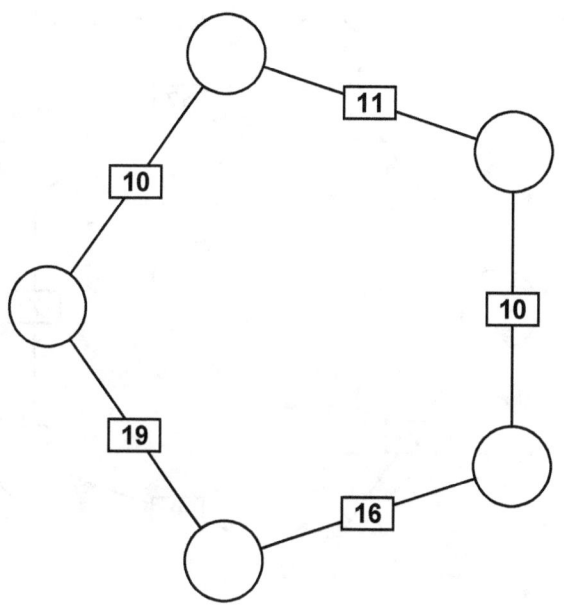

Puzzle 171

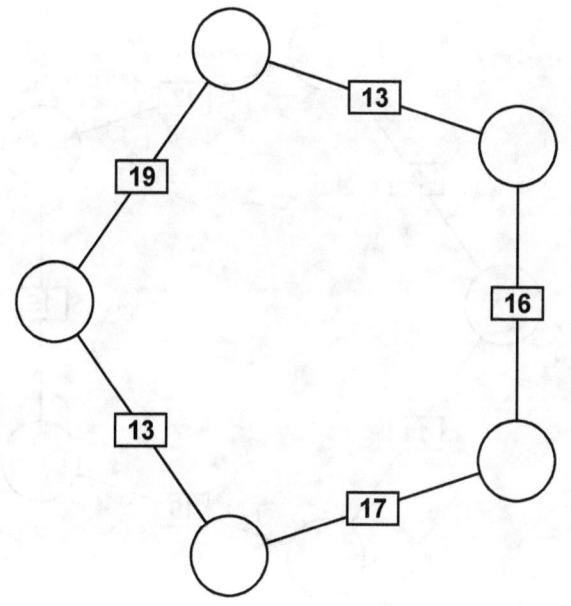

Puzzle 172

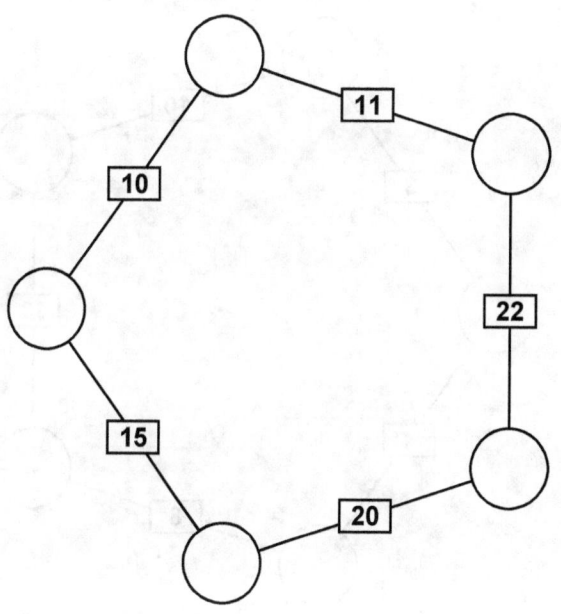

Puzzle 173

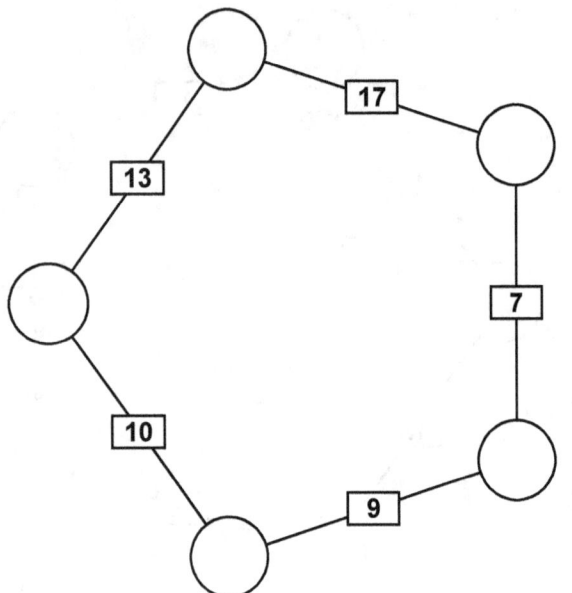

Puzzle 174

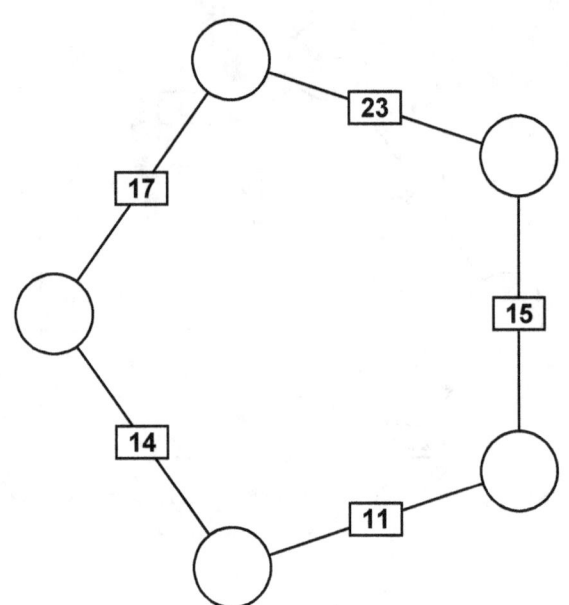

Puzzle 175

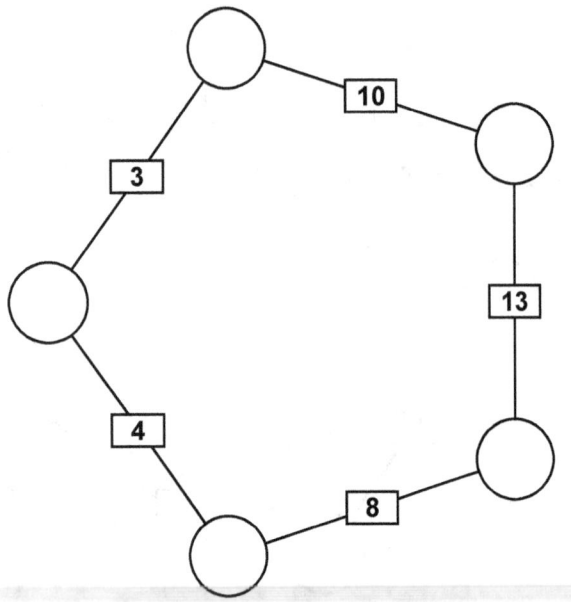

Puzzle 176

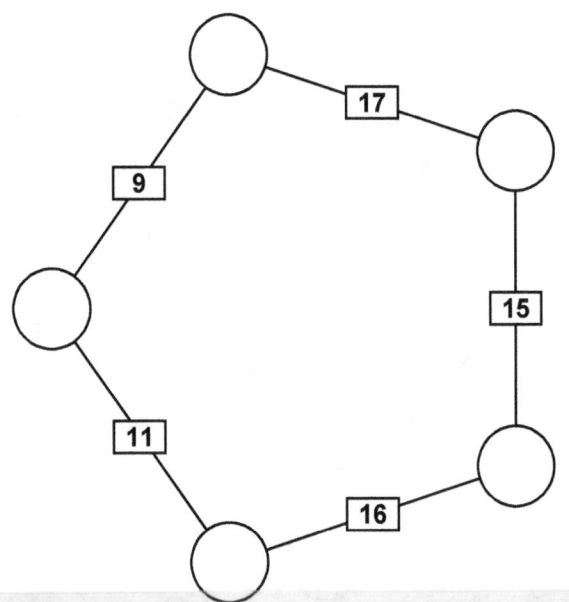

Puzzle 177

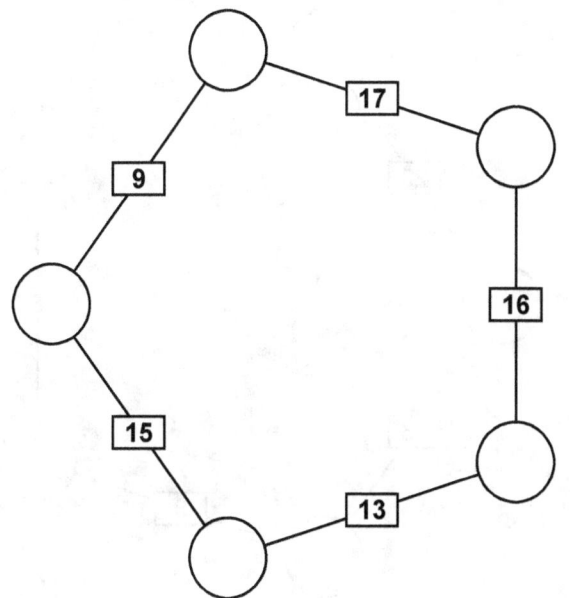

Puzzle 178

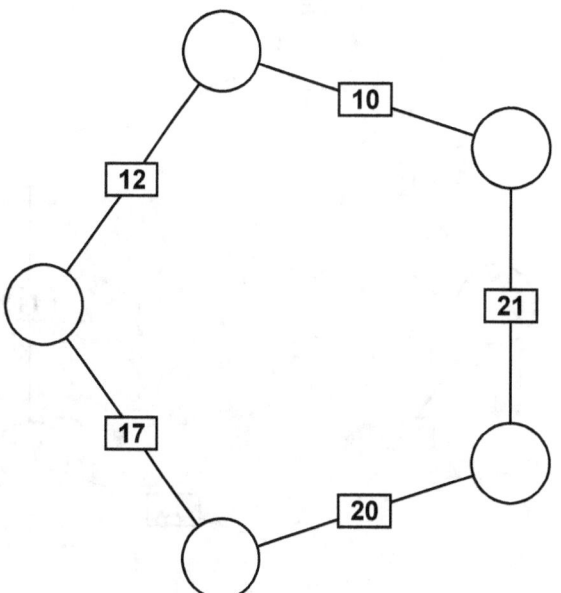

Puzzle 179

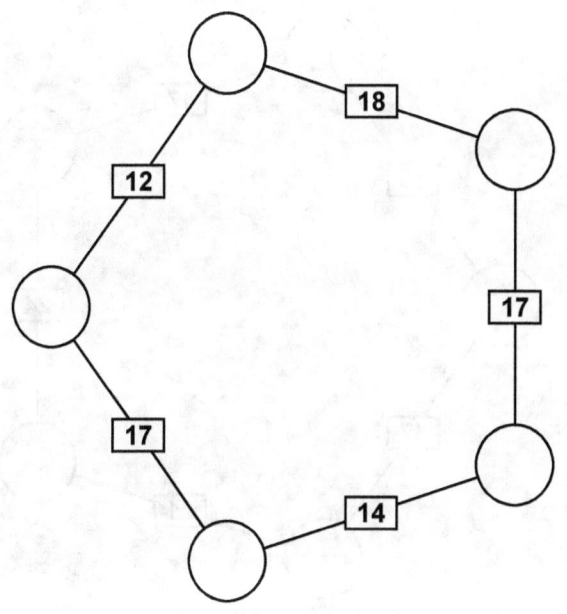

Puzzle 180

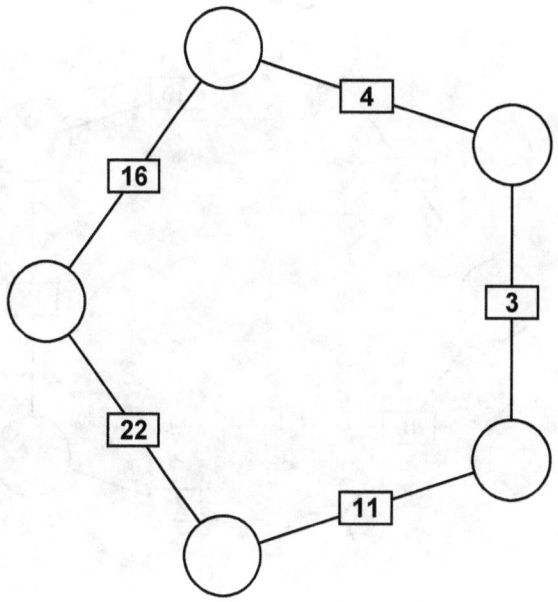

Puzzle 181

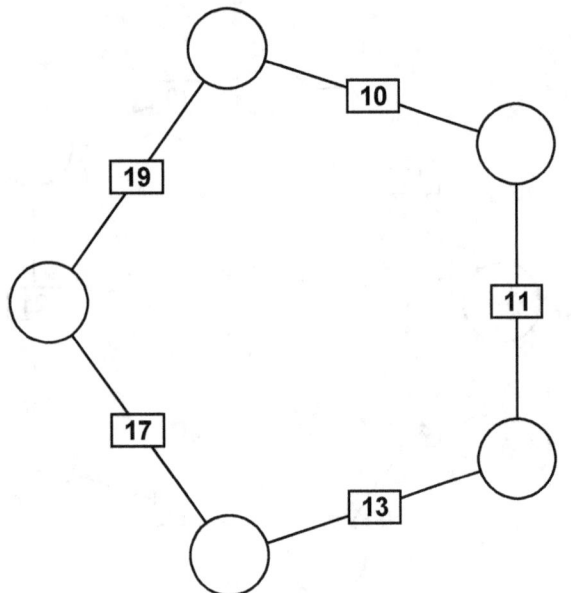

Puzzle 182

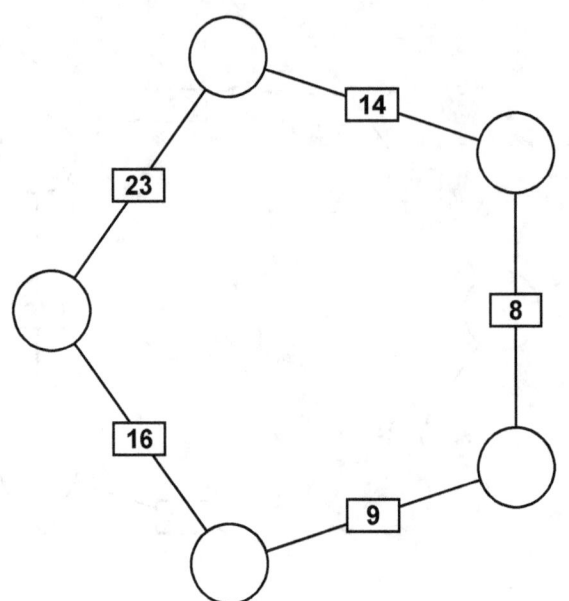

Puzzle 183

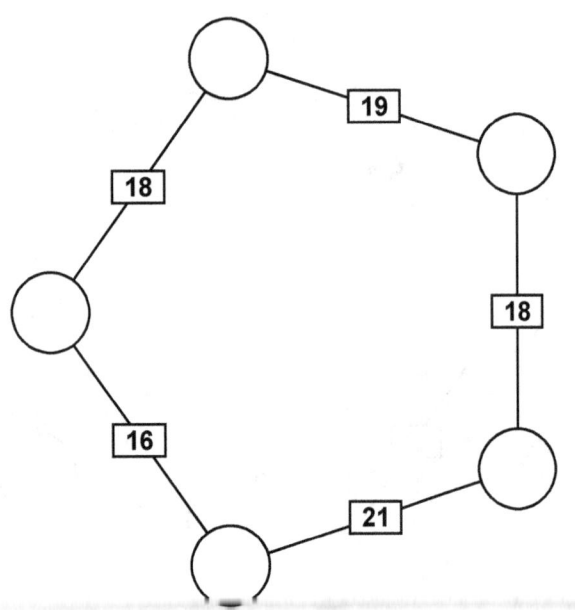

Puzzle 184

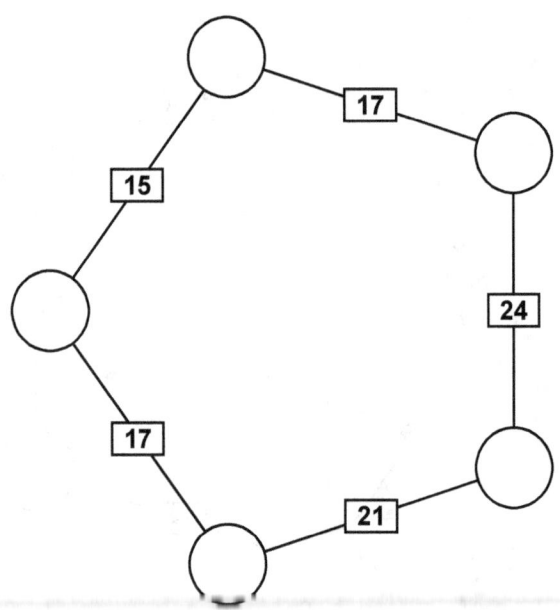

Puzzle 185

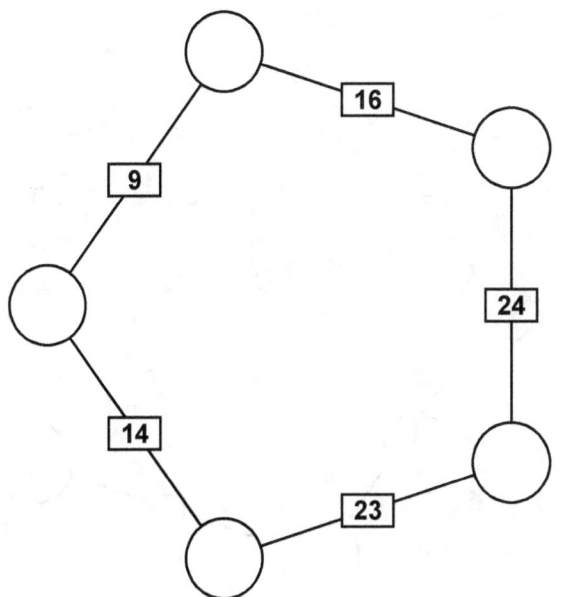

Puzzle 186

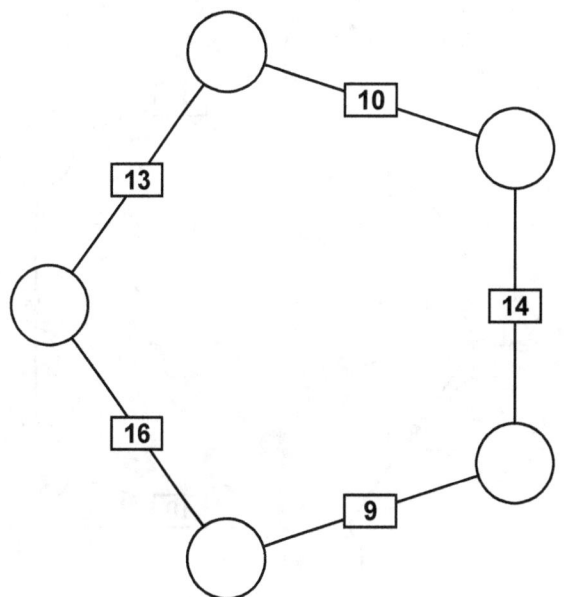

Puzzle 187

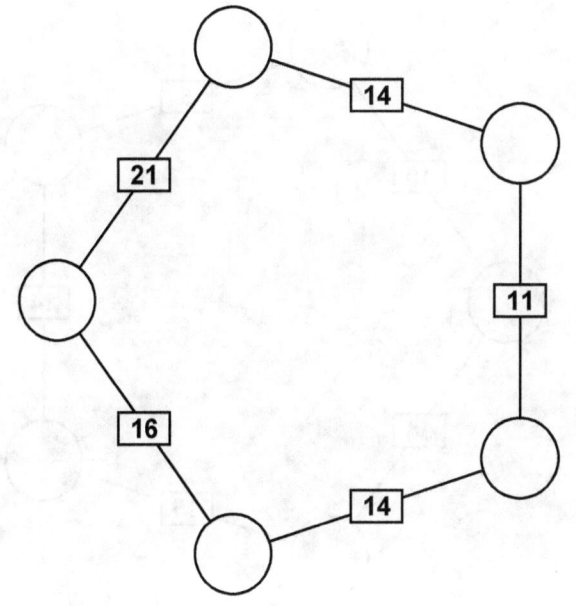

Puzzle 188

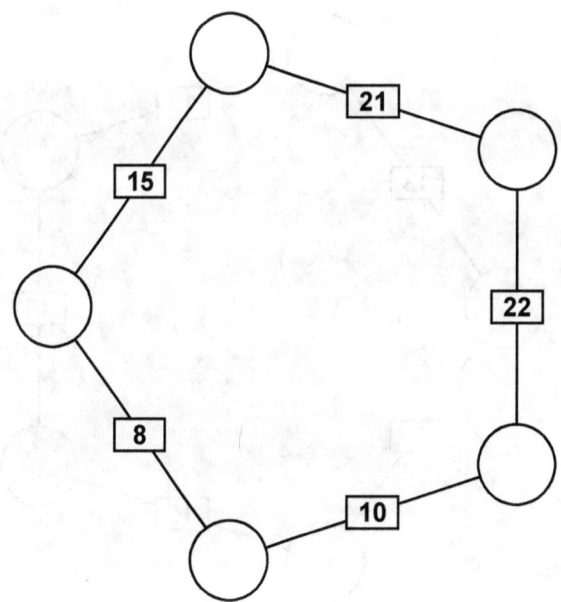

Puzzle 189

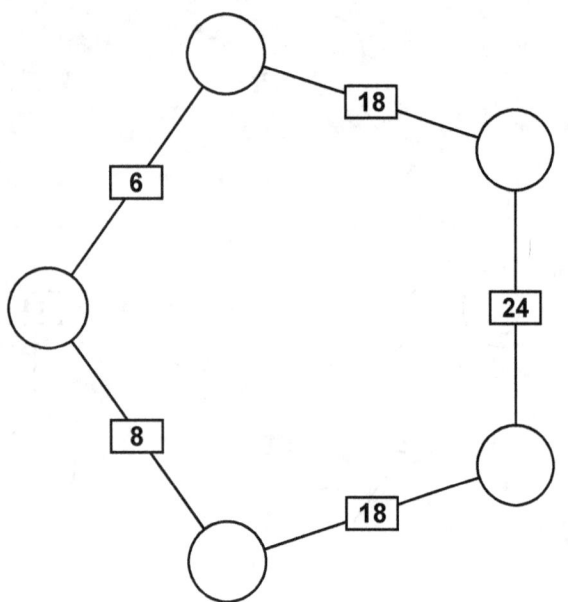

Puzzle 190

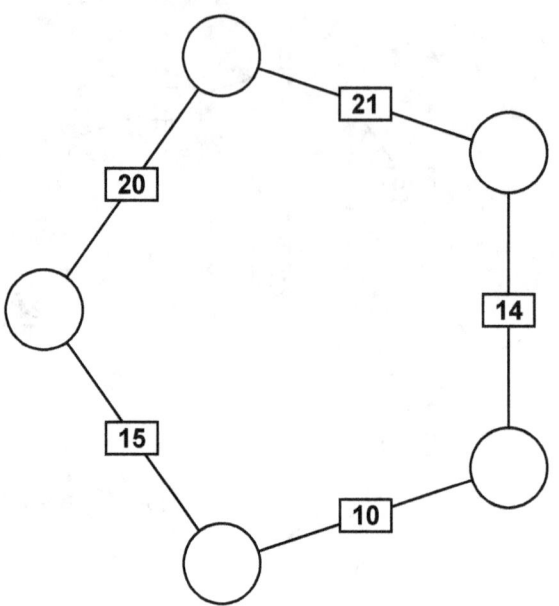

Puzzle 191

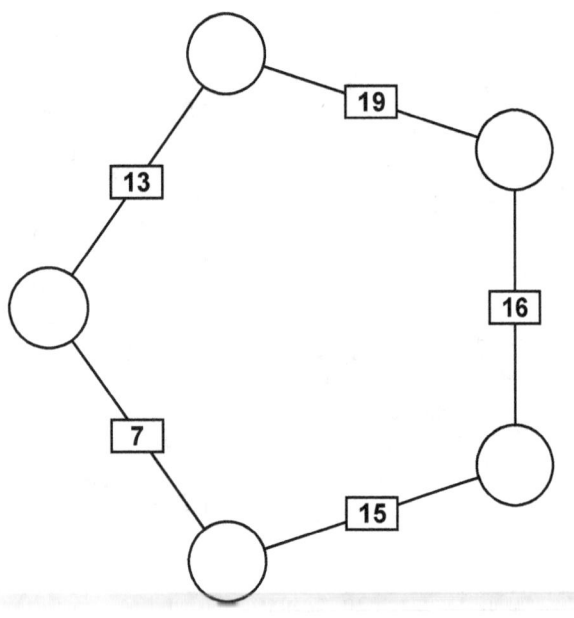

Puzzle 192

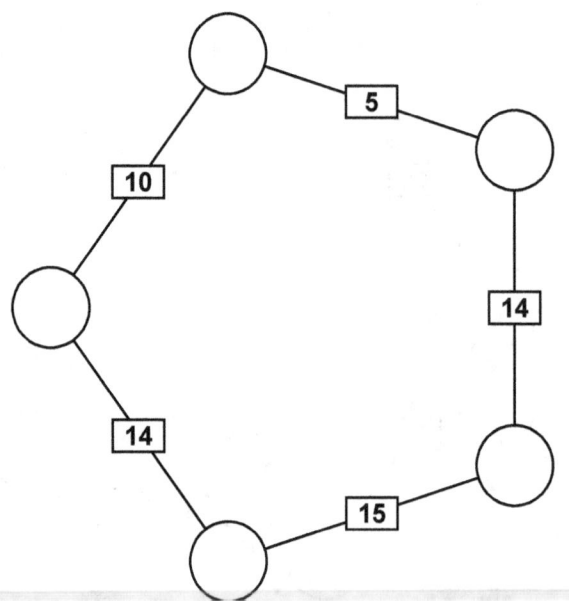

Puzzle 193

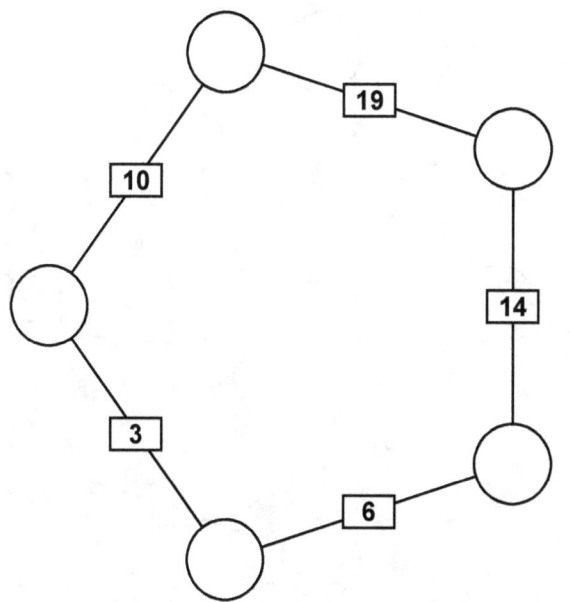

Puzzle 194

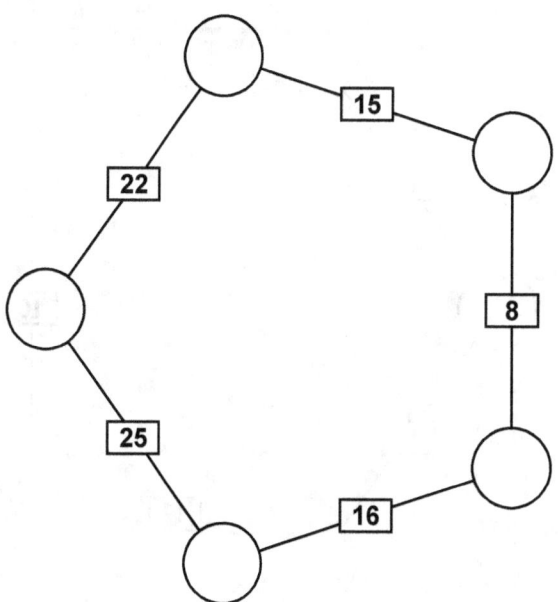

Puzzle 195

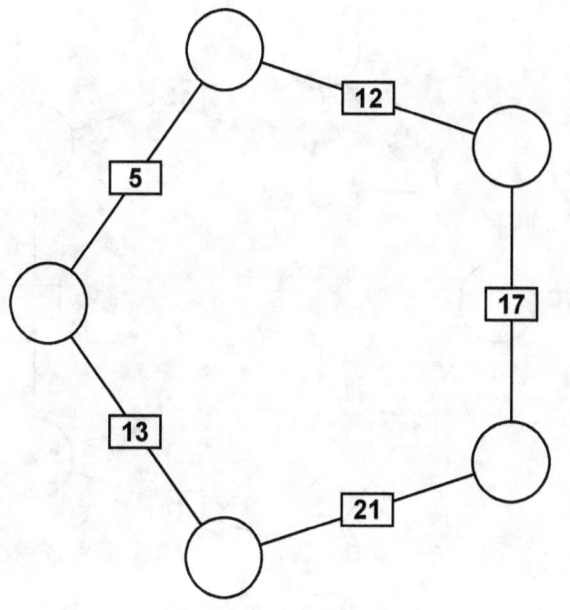

Puzzle 196

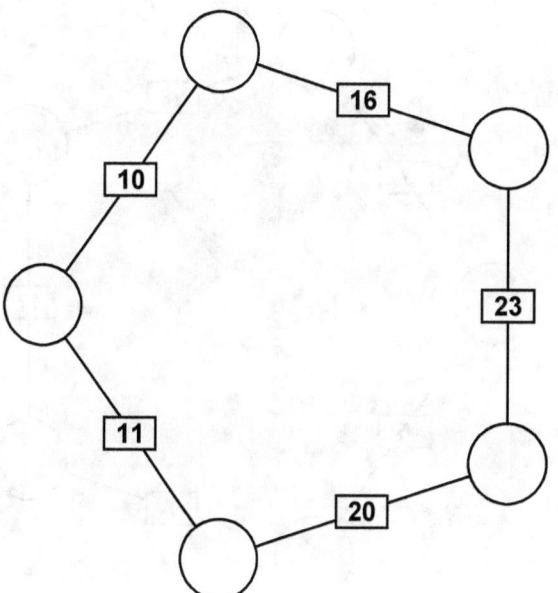

Puzzle 197

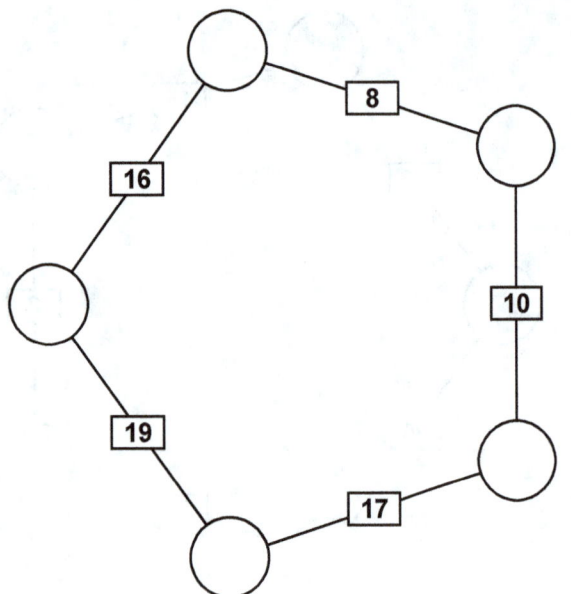

Puzzle 198

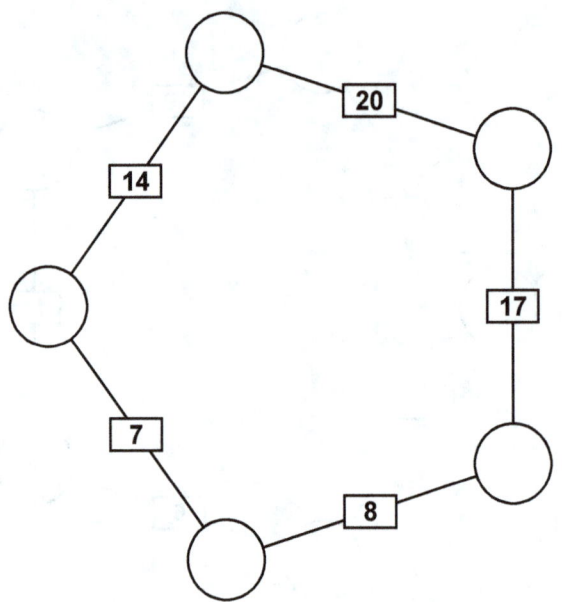

Puzzle 199

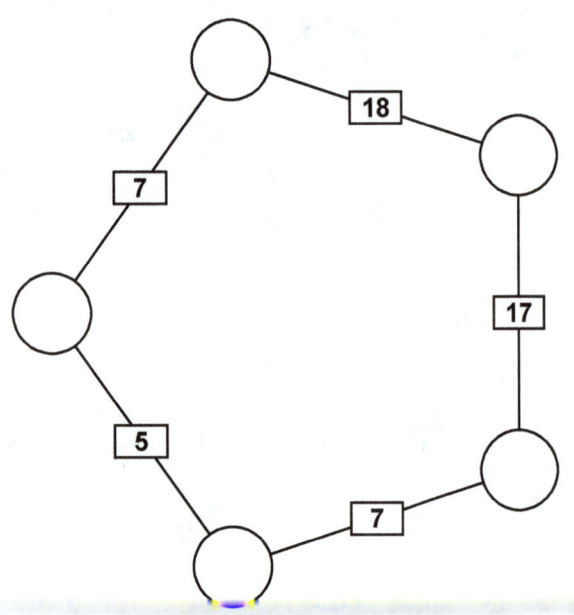

Puzzle 200

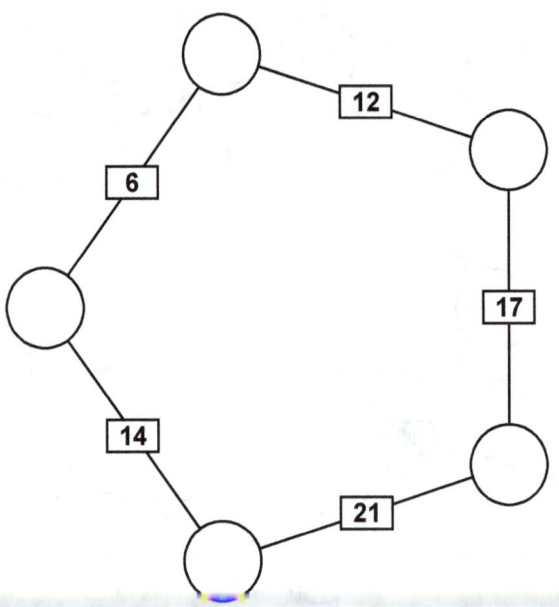

Puzzle 201

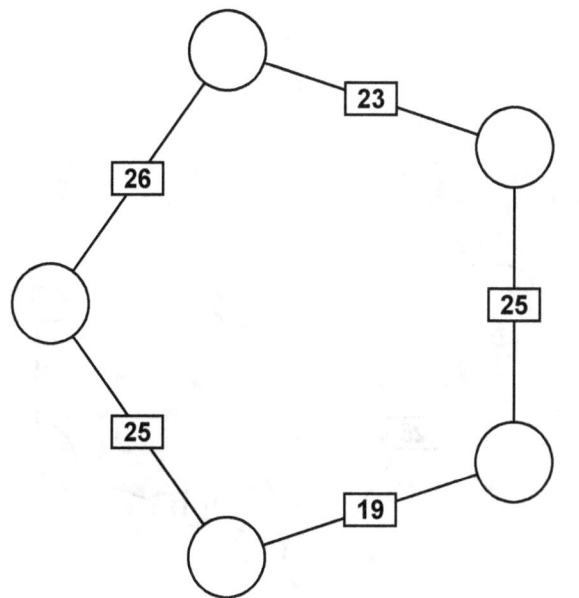

Puzzle 202

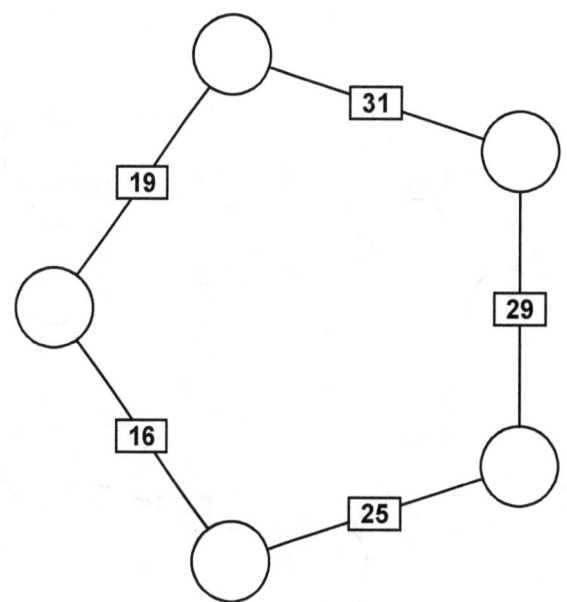

Puzzle 203

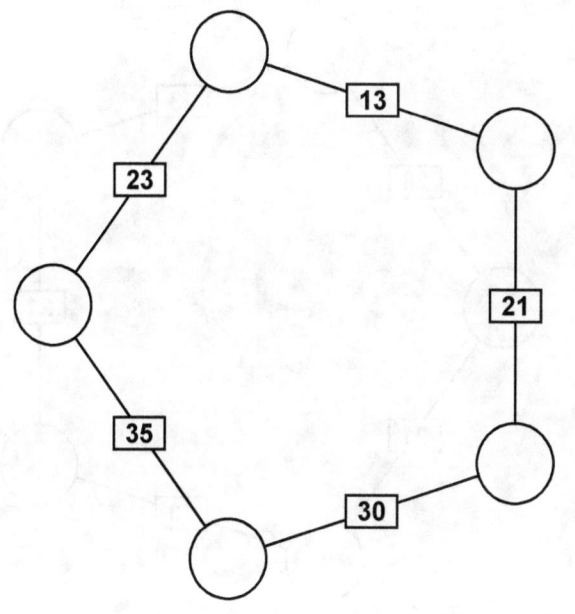

Puzzle 204

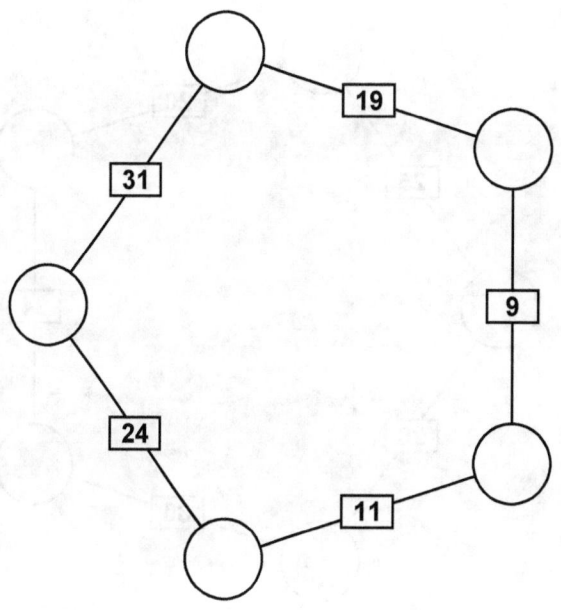

Puzzle 205

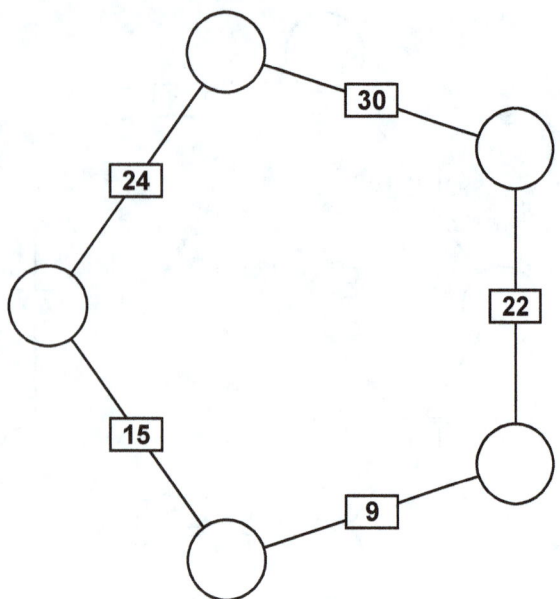

Puzzle 206

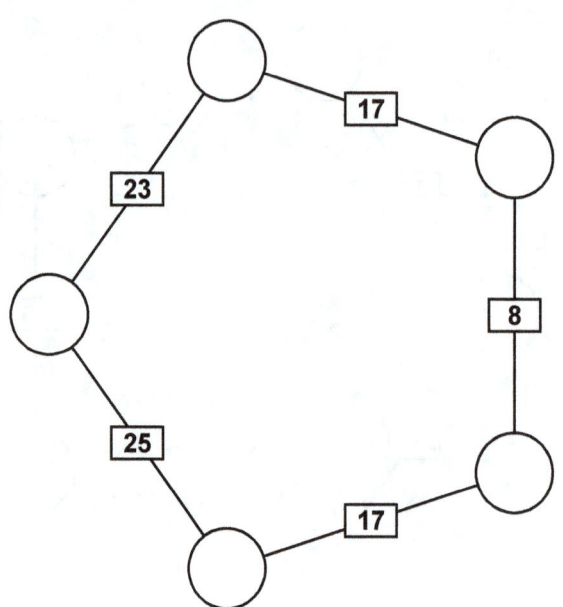

Puzzle 207

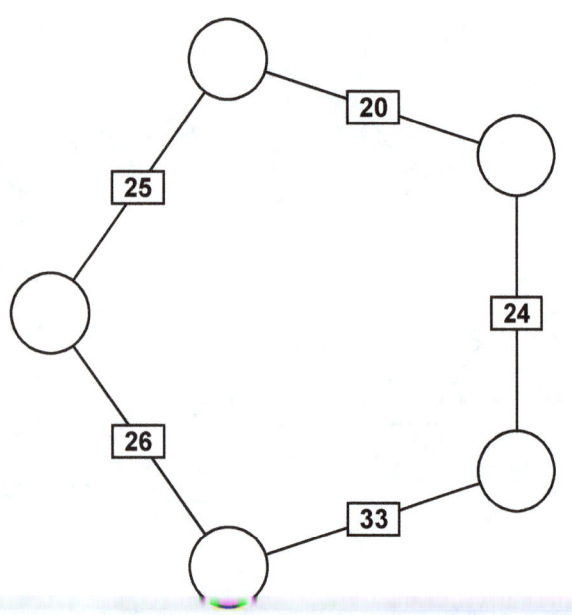

Puzzle 208

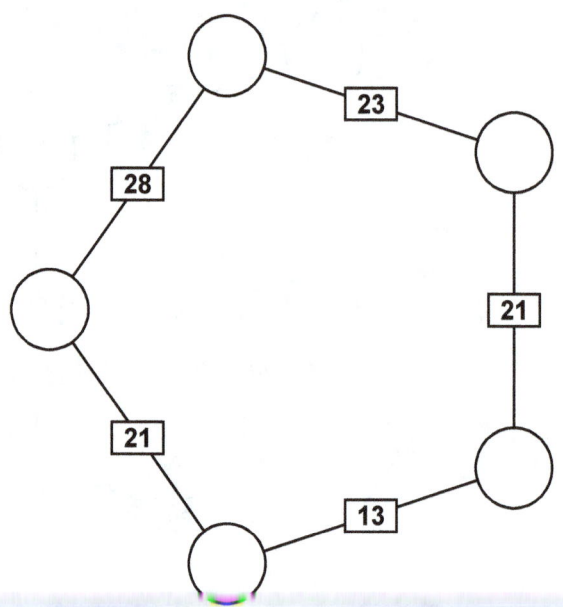

Puzzle 209

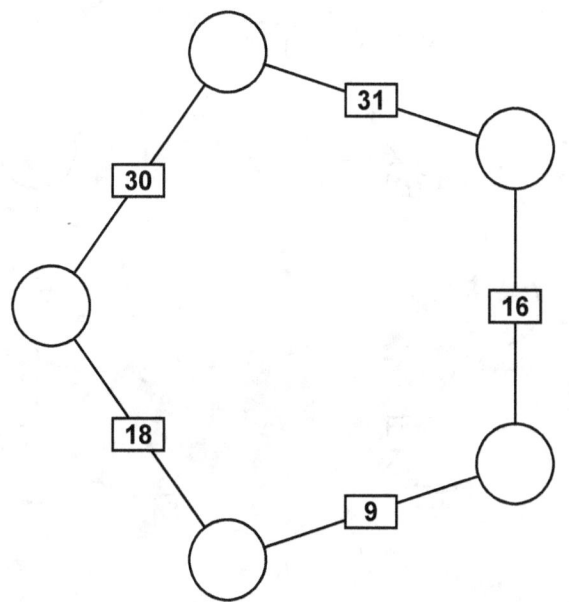

Puzzle 210

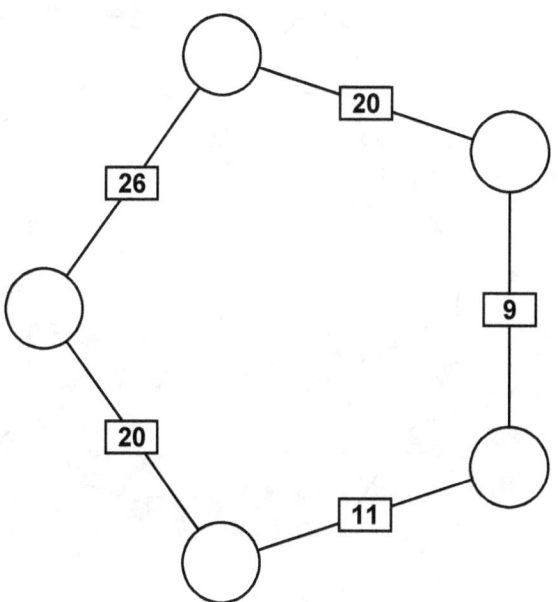

Puzzle 211

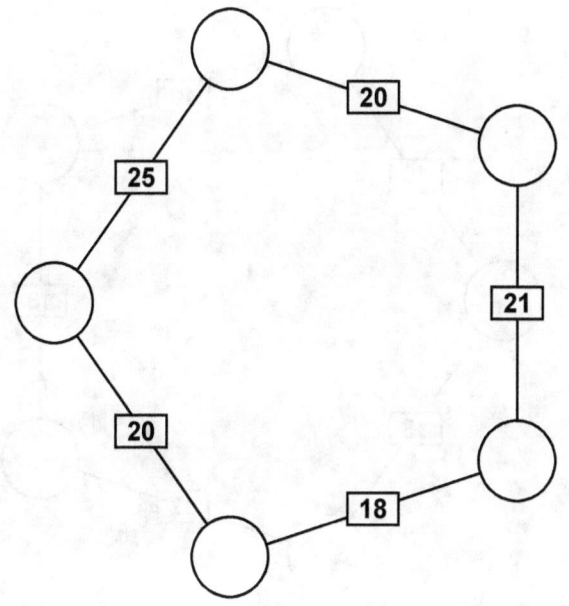

Puzzle 212

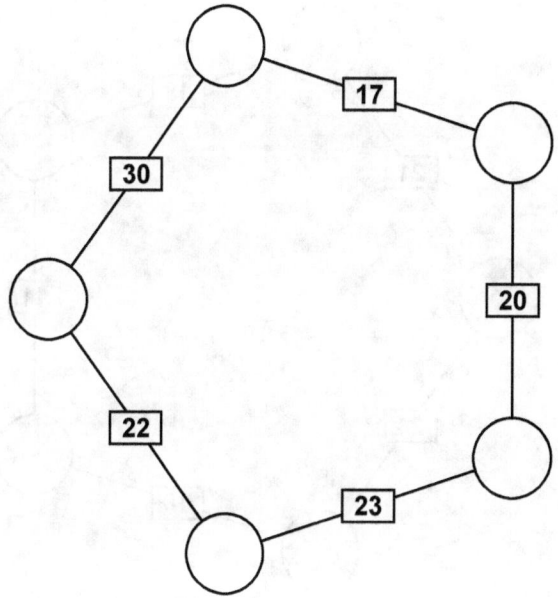

Puzzle 213

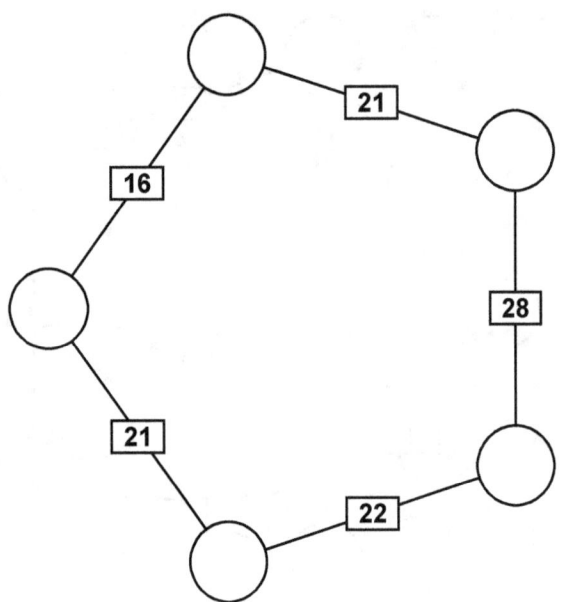

Puzzle 214

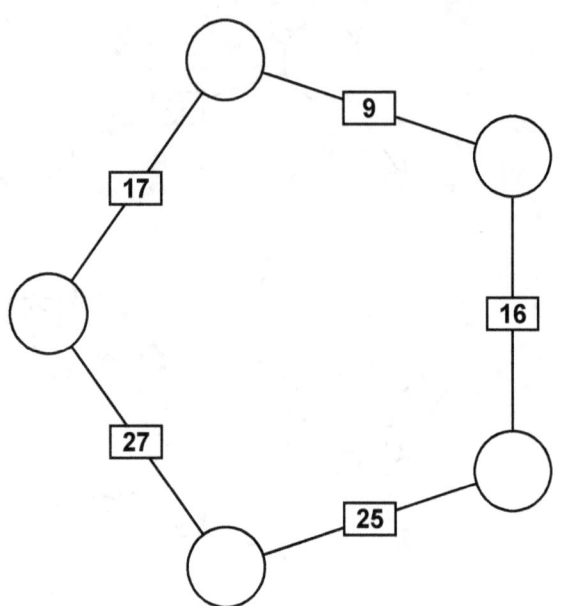

Puzzle 215

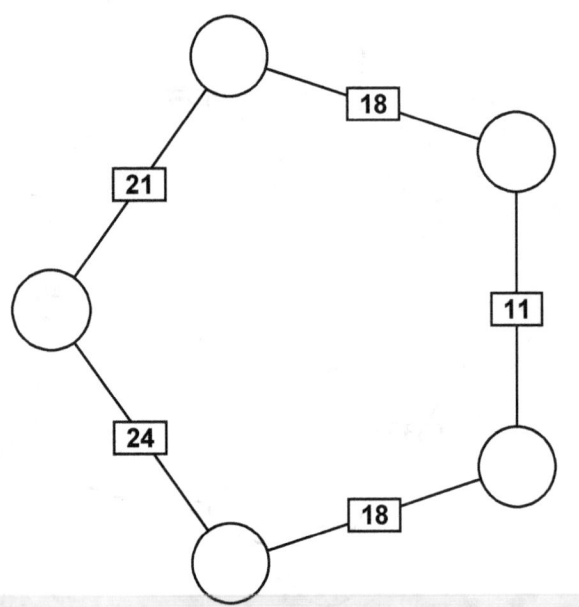

Puzzle 216

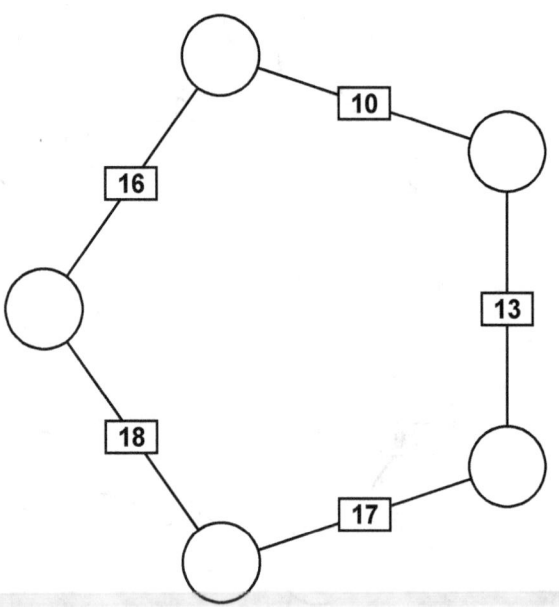

Puzzle 217

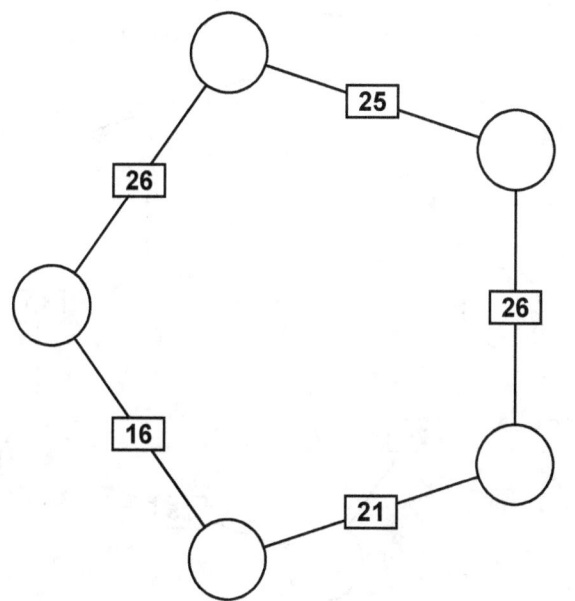

Puzzle 218

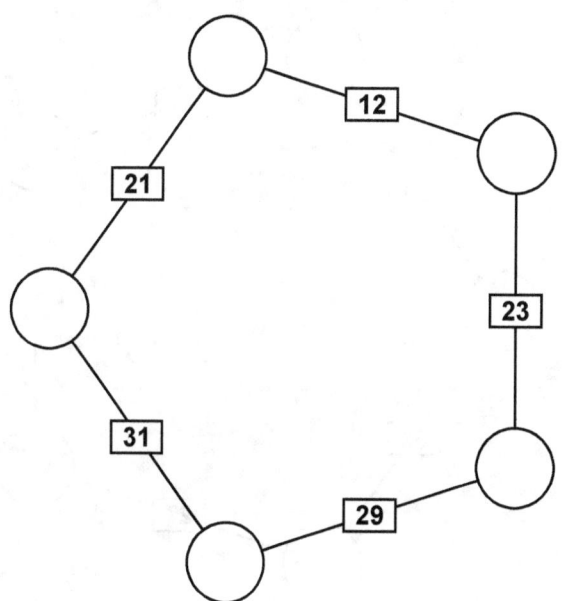

Puzzle 219

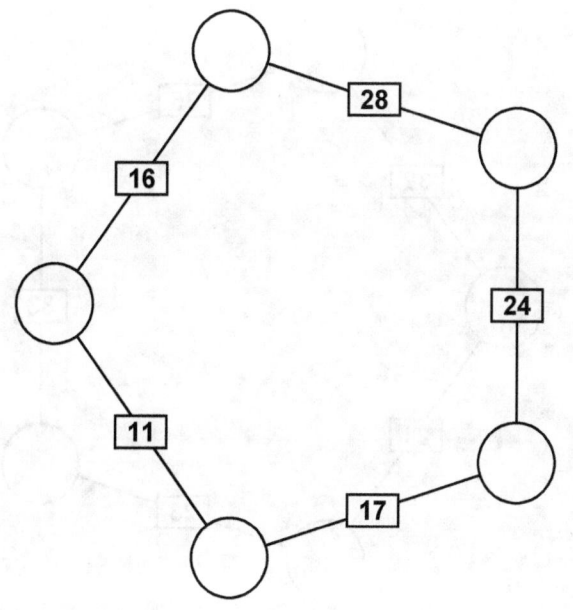

Puzzle 220

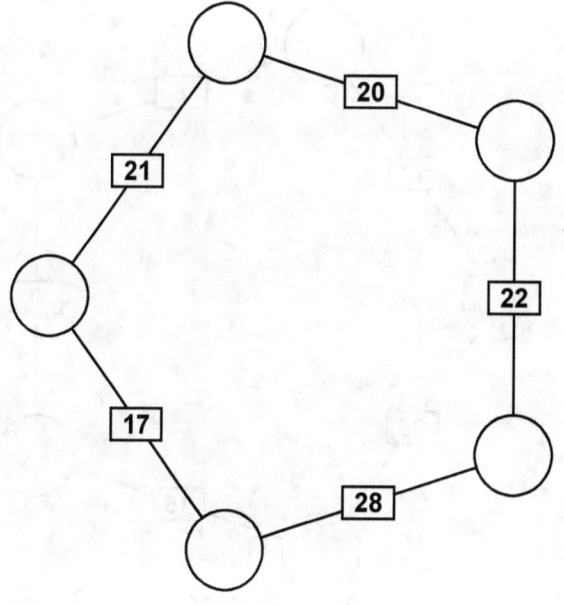

Puzzle 221

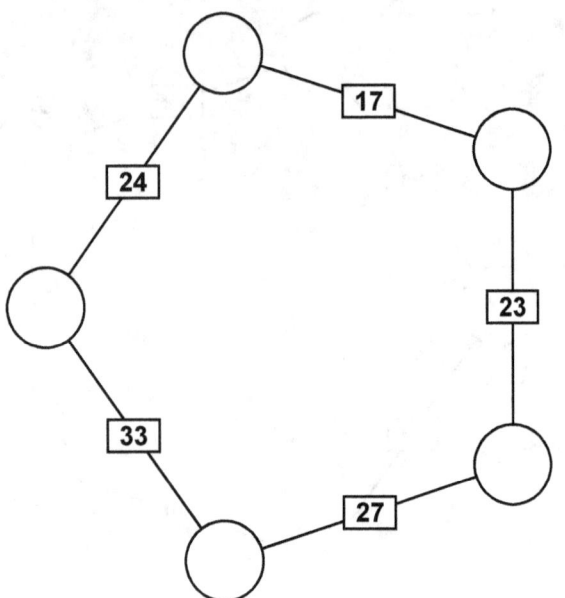

Puzzle 222

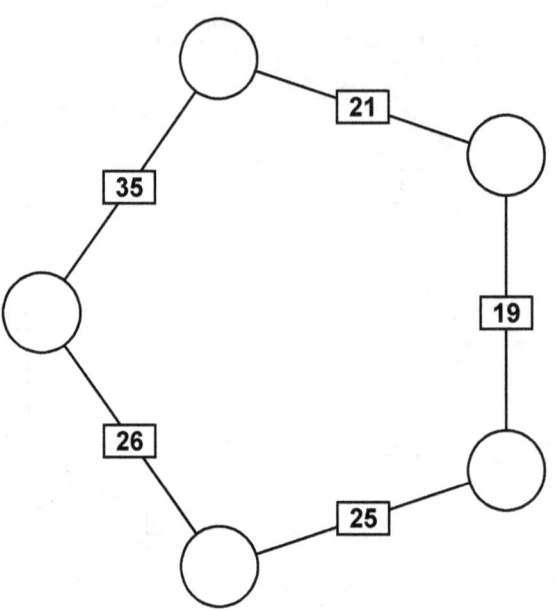

Puzzle 223

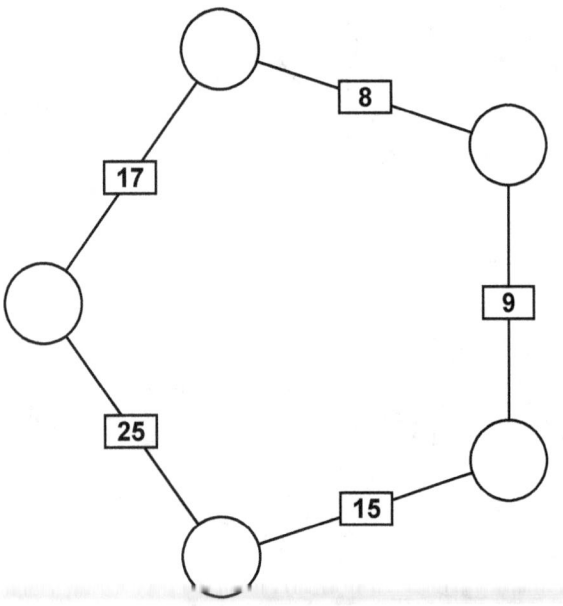

Puzzle 224

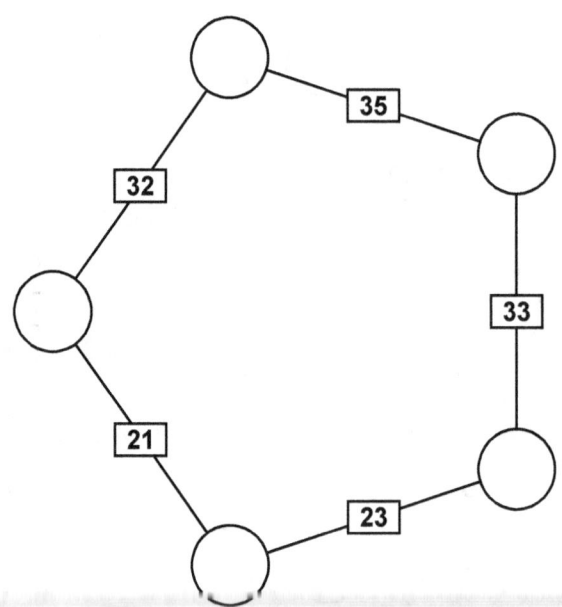

Puzzle 225

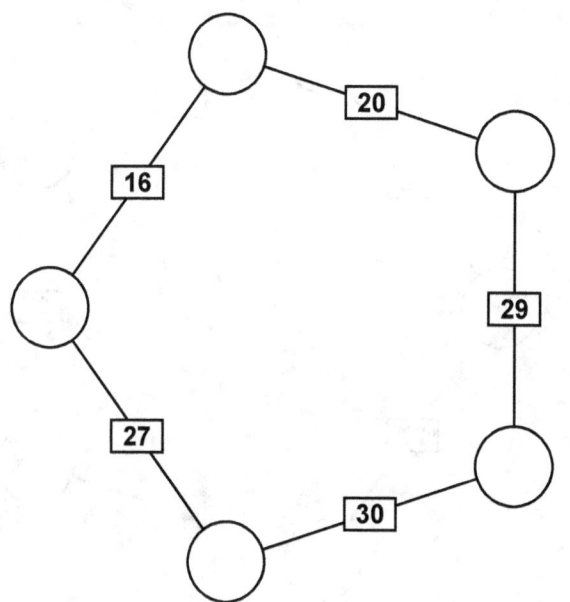

Puzzle 226

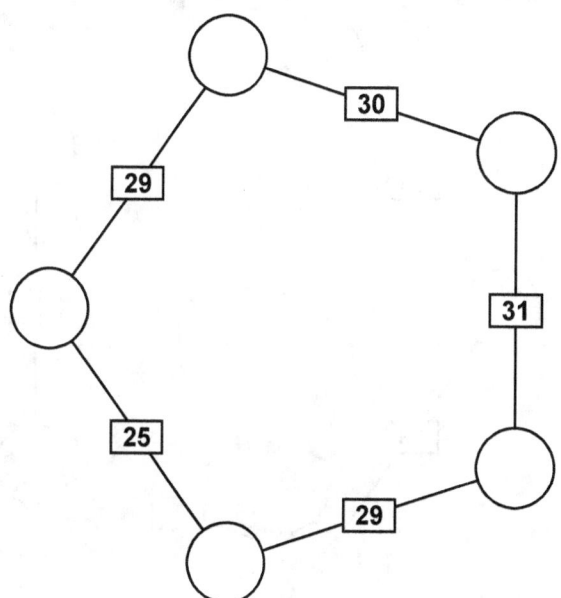

Puzzle 227

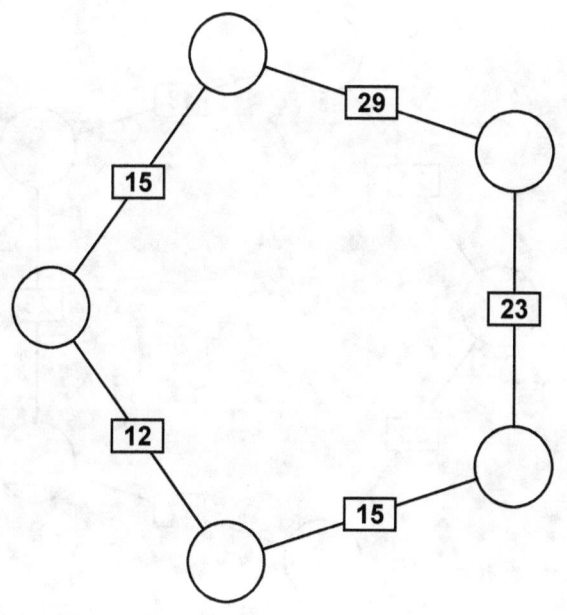

Puzzle 228

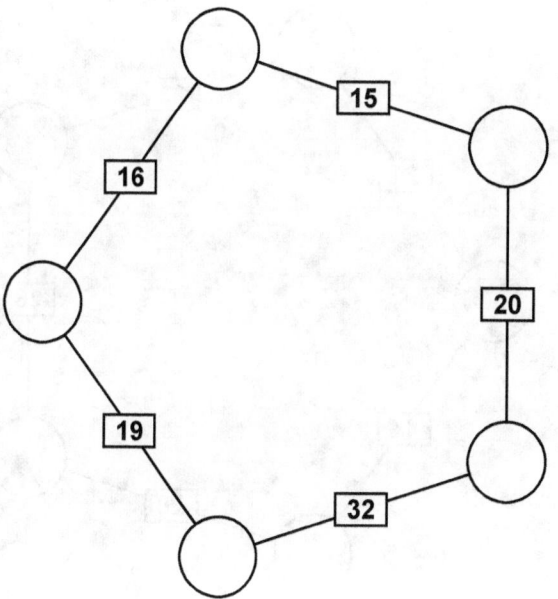

Puzzle 229

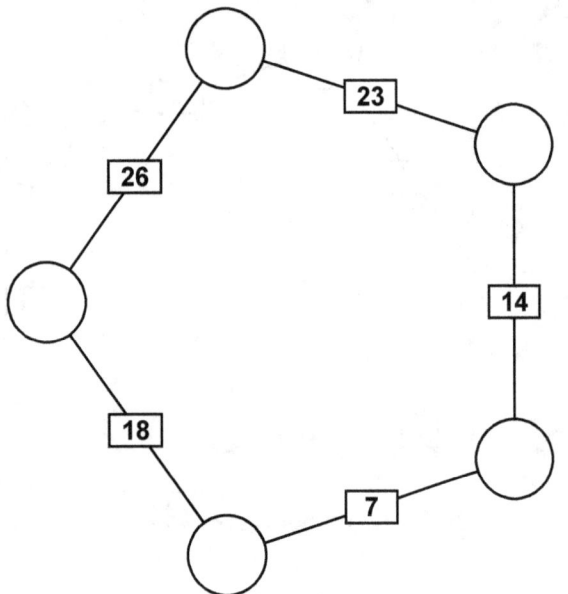

Puzzle 230

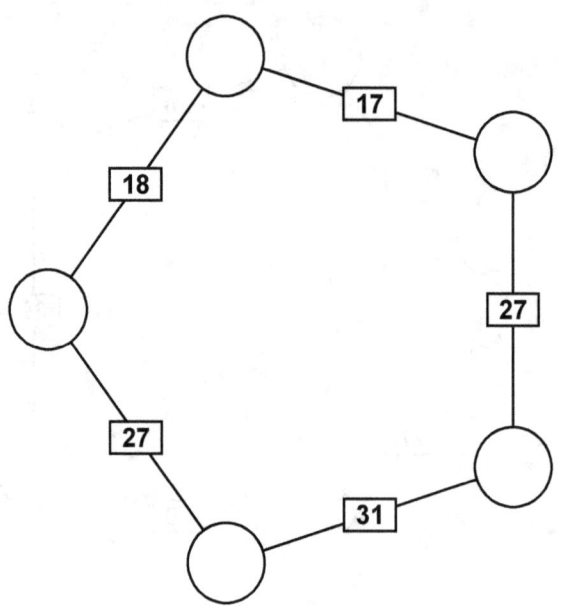

Puzzle 231

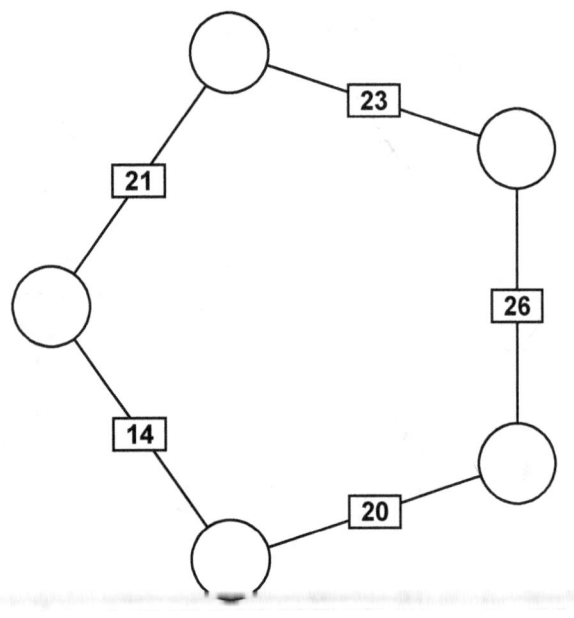

Puzzle 232

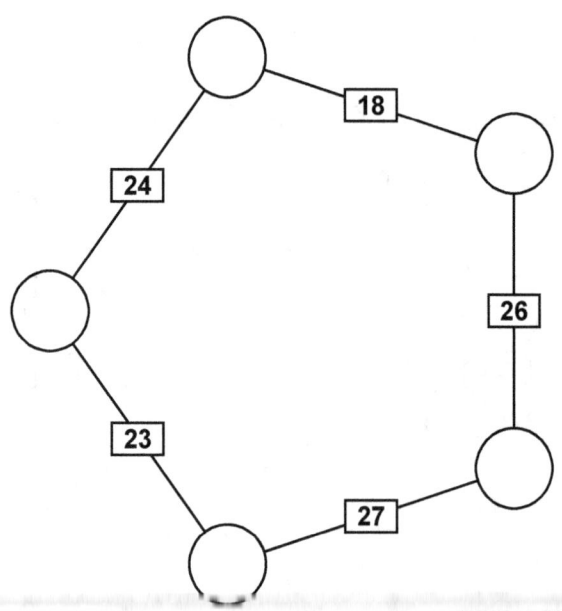

Puzzle 233

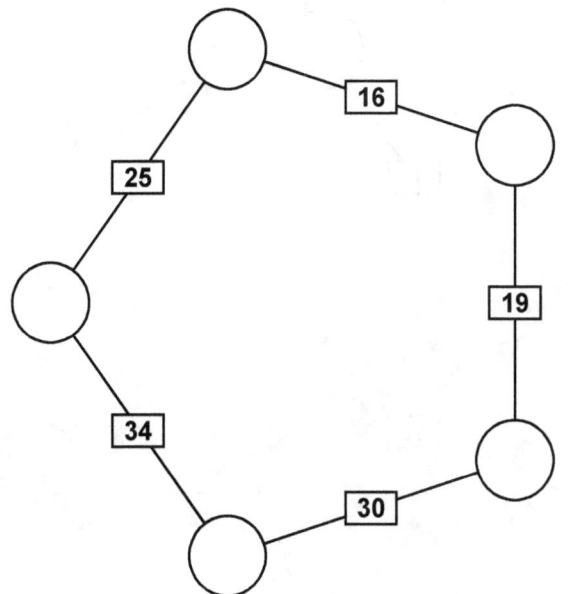

Puzzle 234

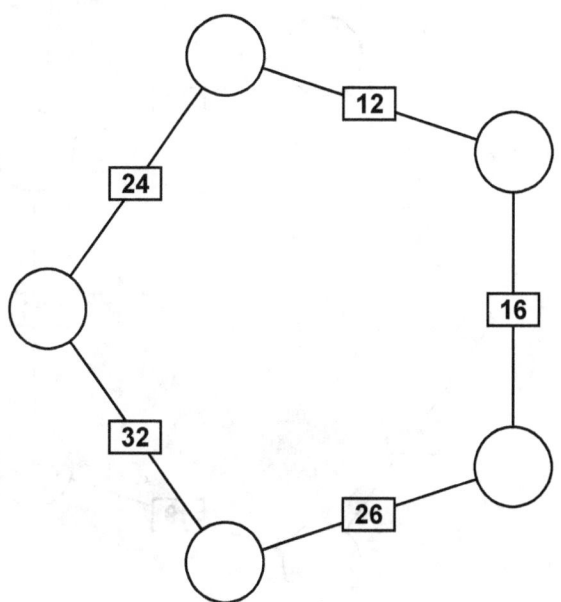

Puzzle 235

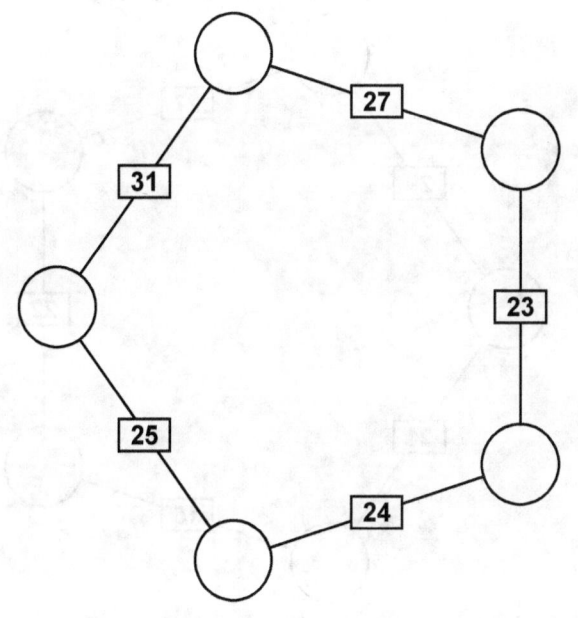

Puzzle 236

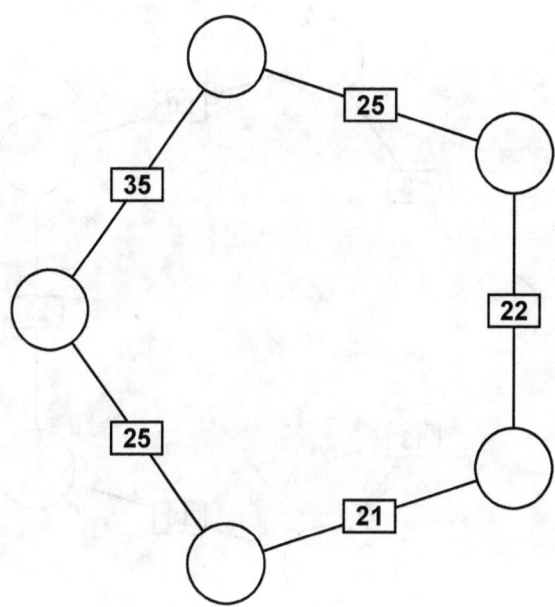

Puzzle 237

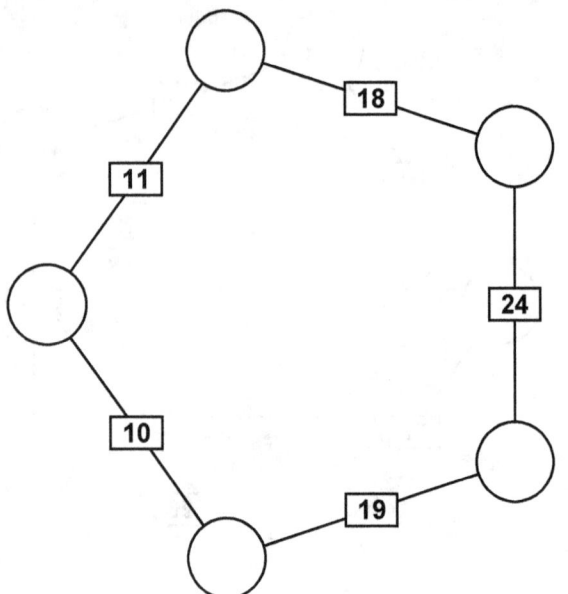

Puzzle 238

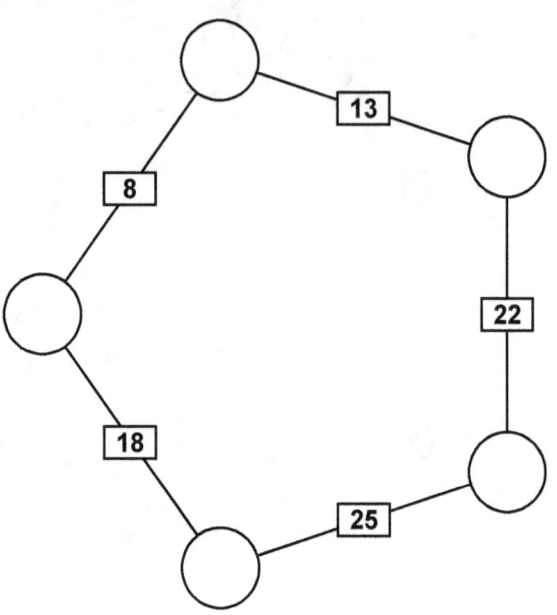

Puzzle 239

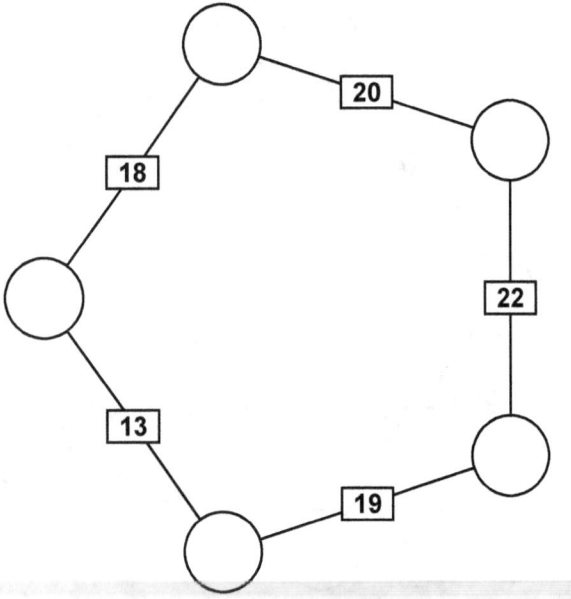

Puzzle 240

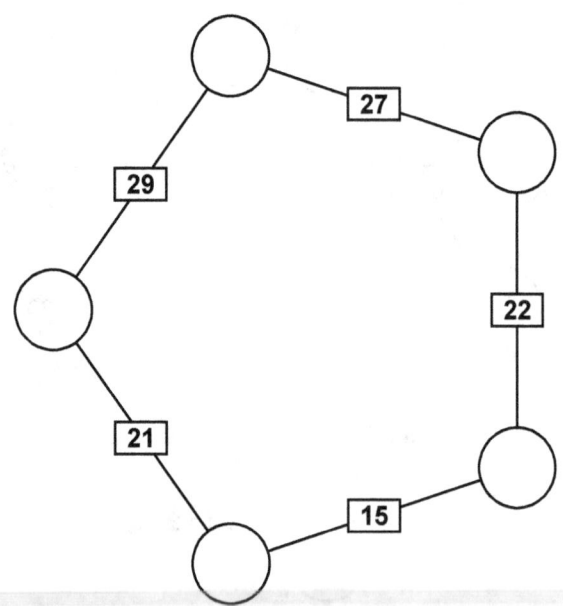

Puzzle 241

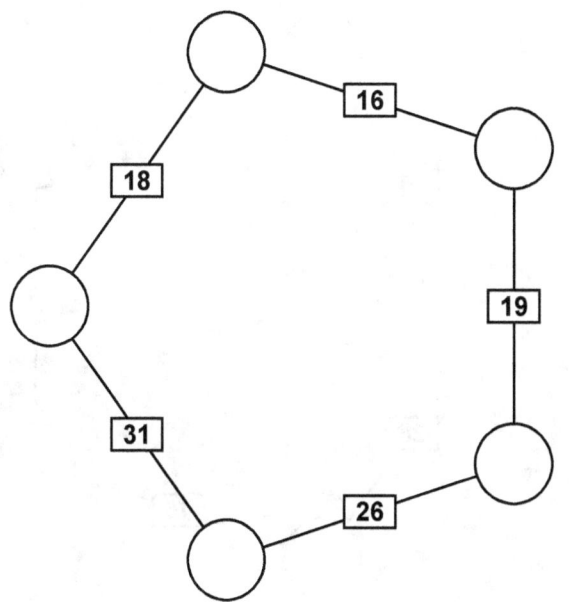

Puzzle 242

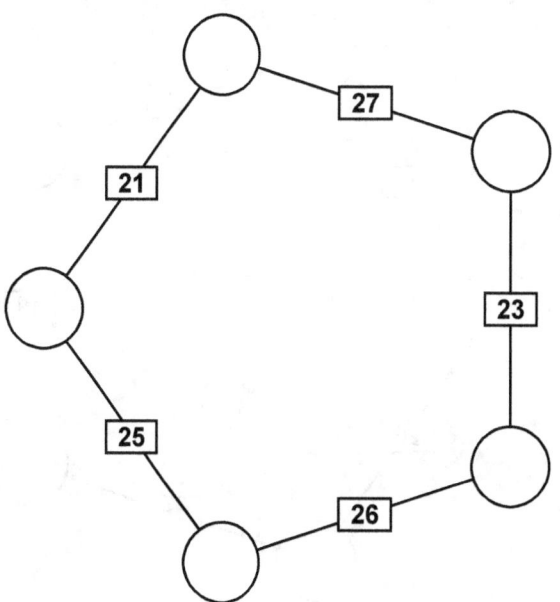

Puzzle 243

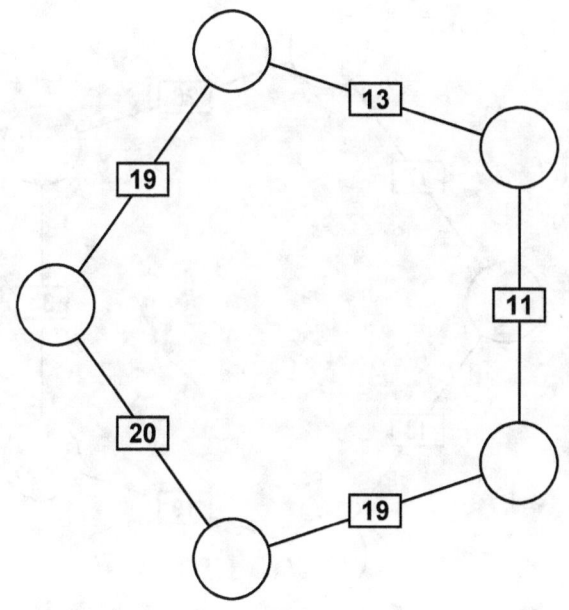

Puzzle 244

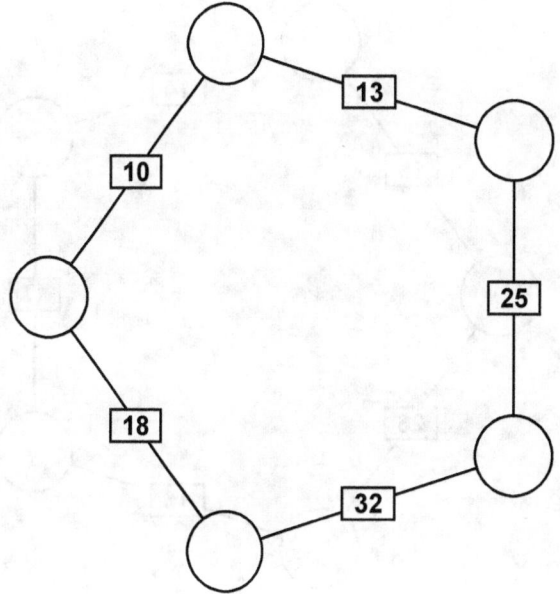

Puzzle 245

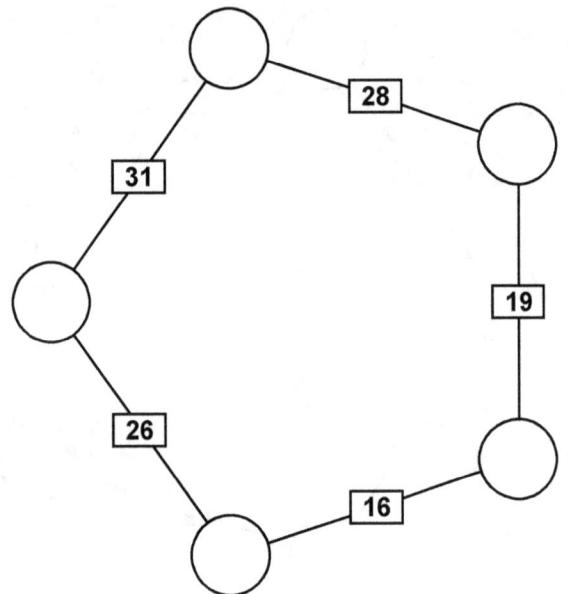

Puzzle 246

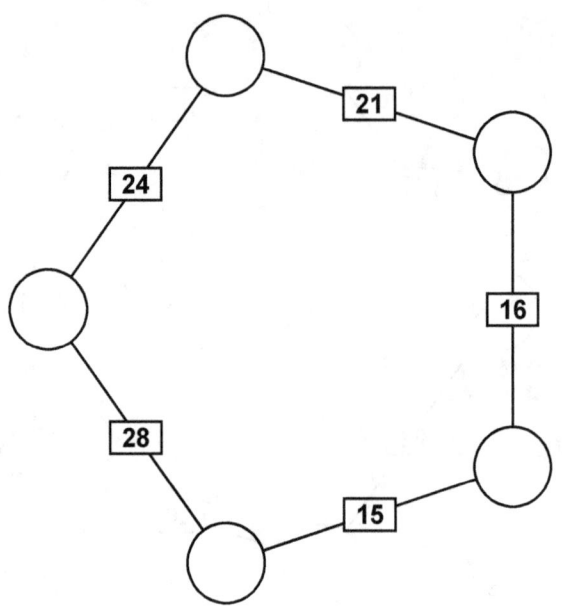

Puzzle 247

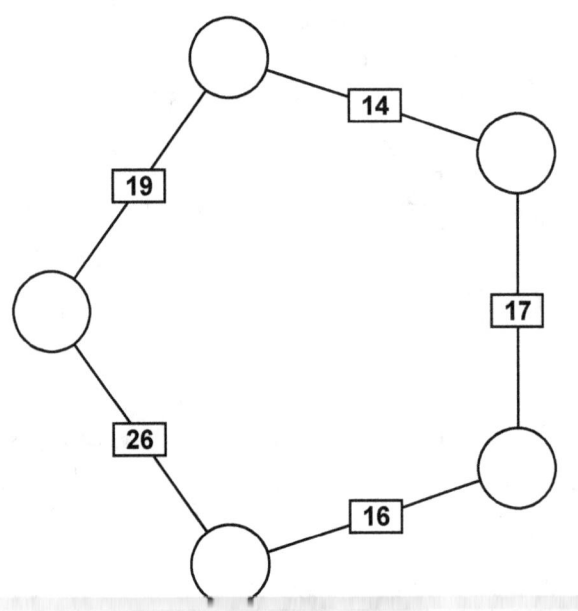

Puzzle 248

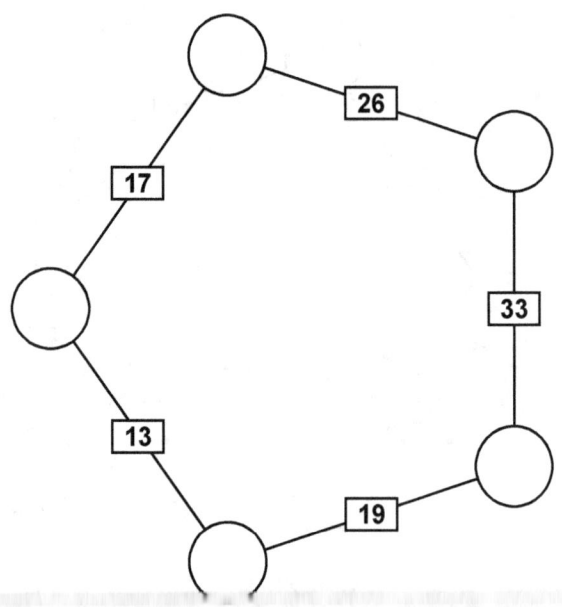

Puzzle 249

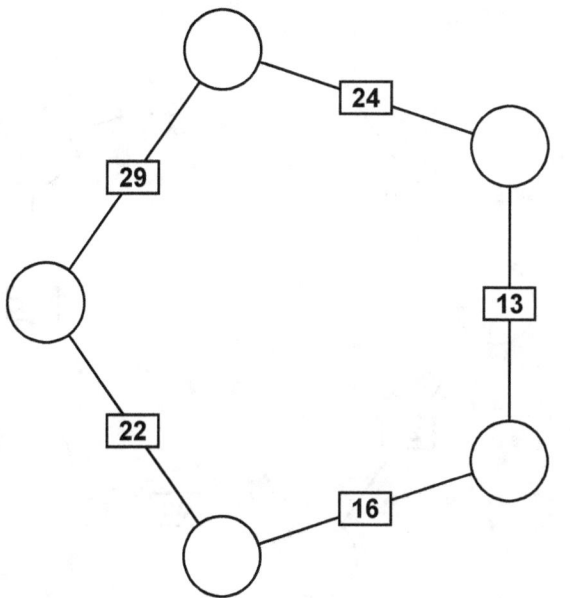

Puzzle 250

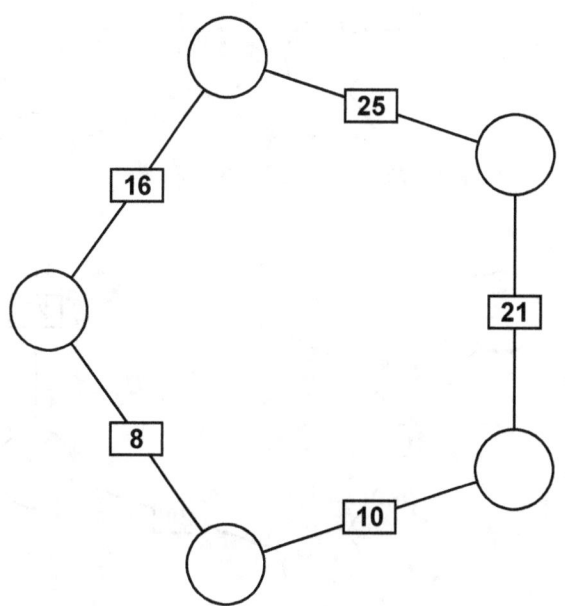

Puzzle 251

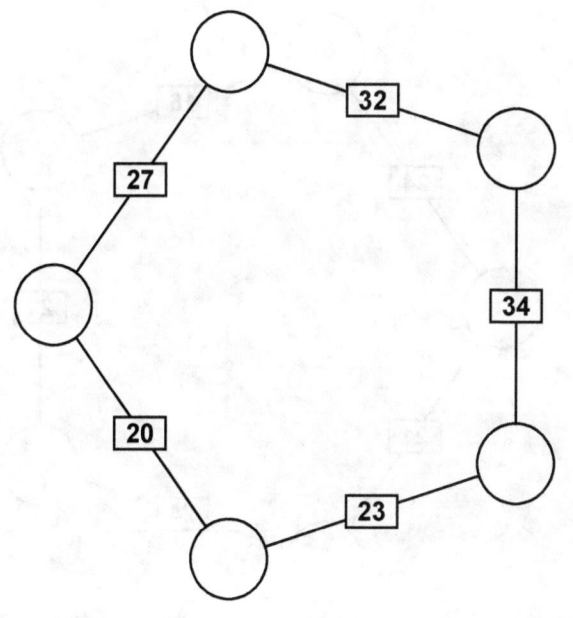

Puzzle 252

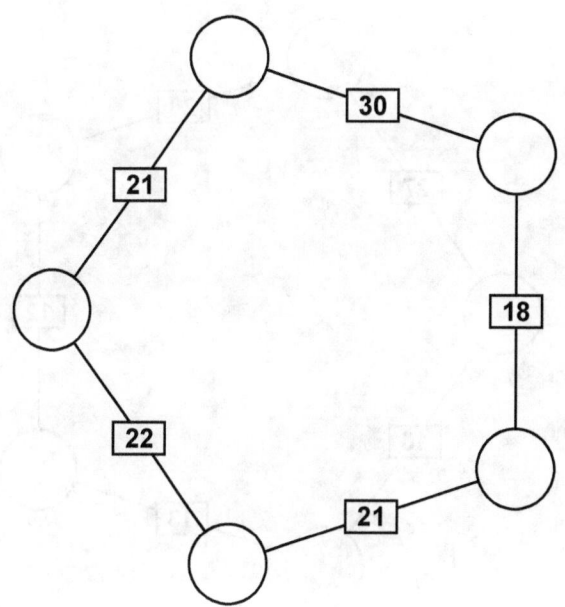

Puzzle 253

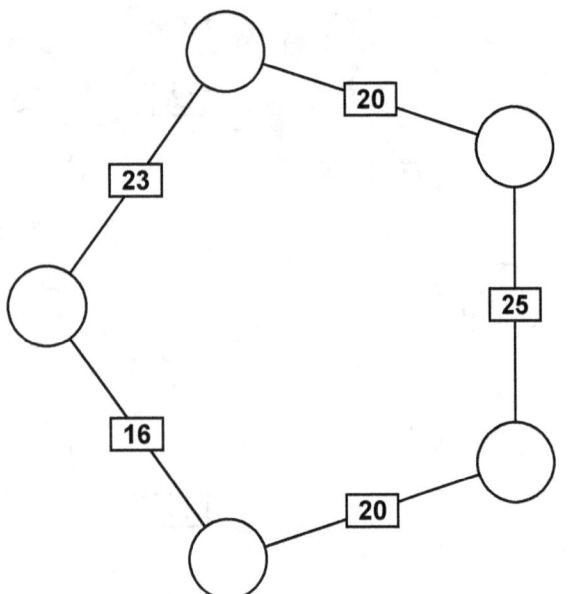

Puzzle 254

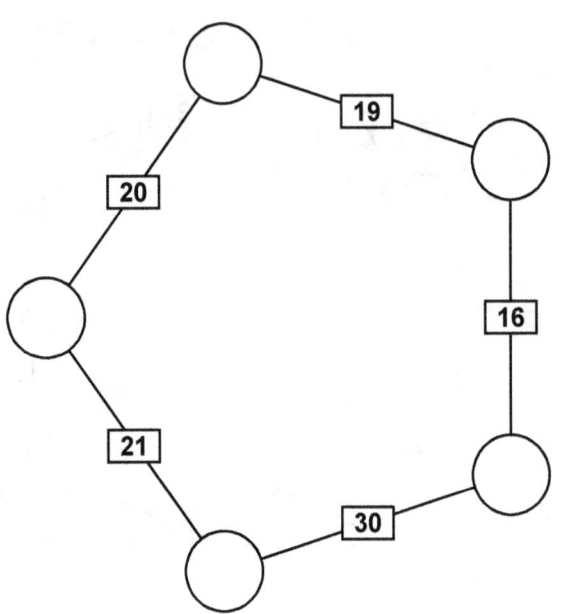

Puzzle 255

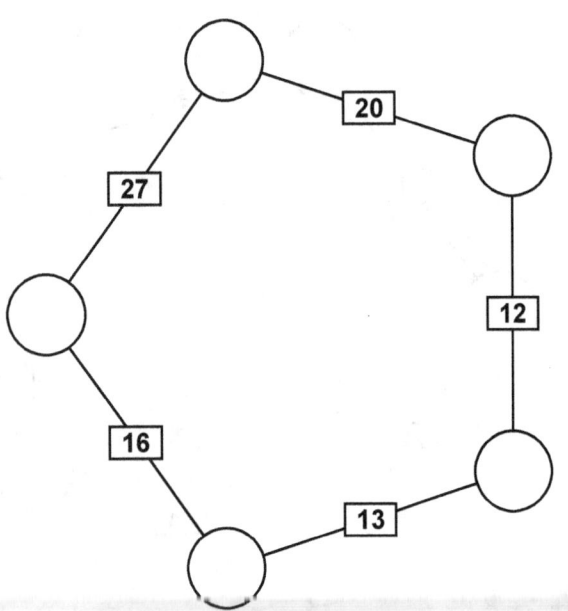

Puzzle 256

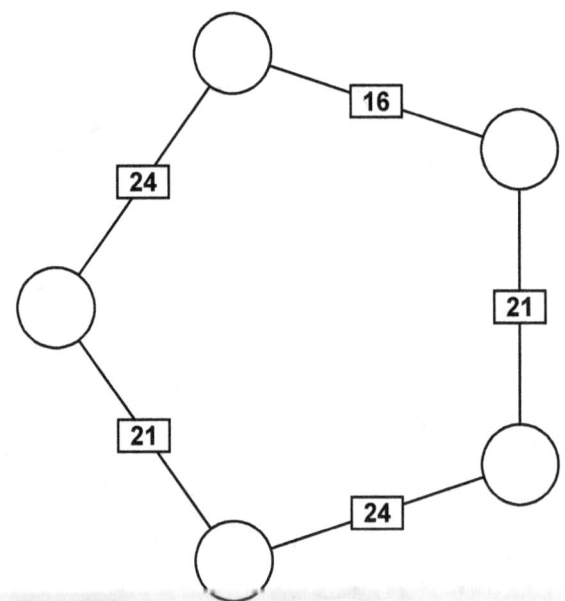

Puzzle 257

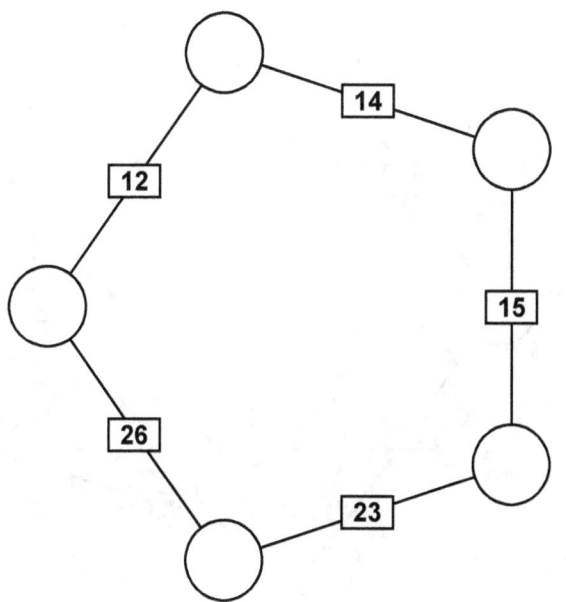

Puzzle 258

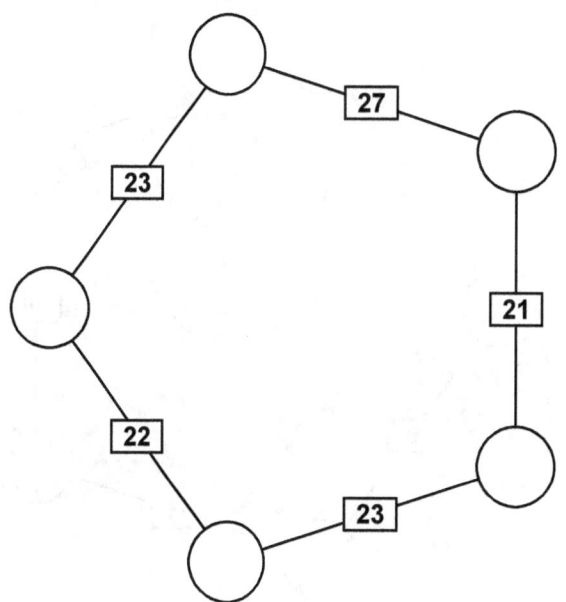

Puzzle 259

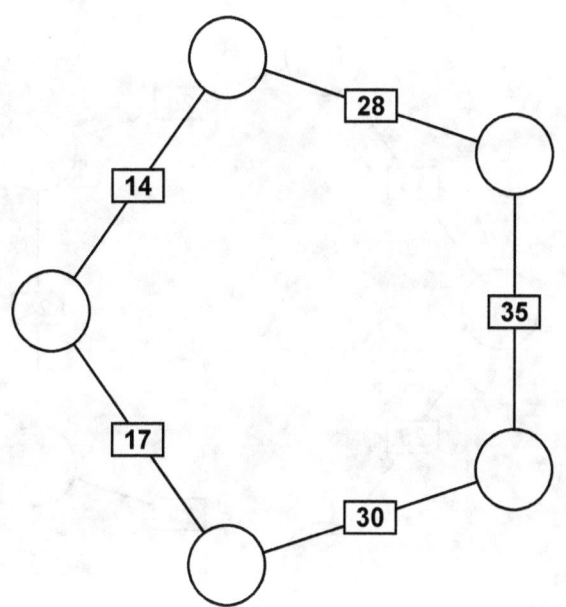

Puzzle 260

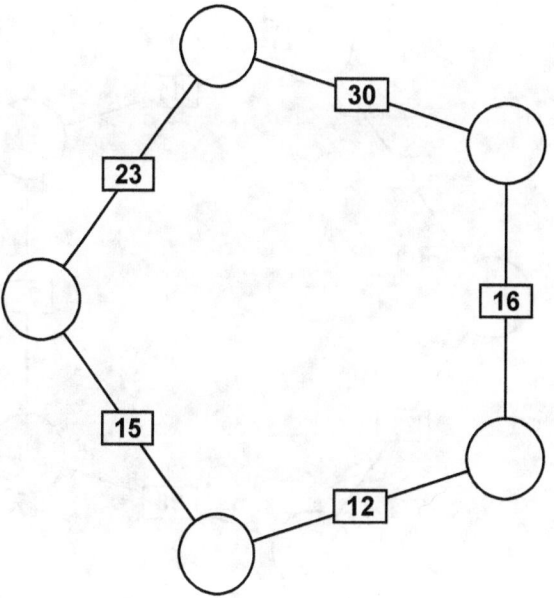

Puzzle 261

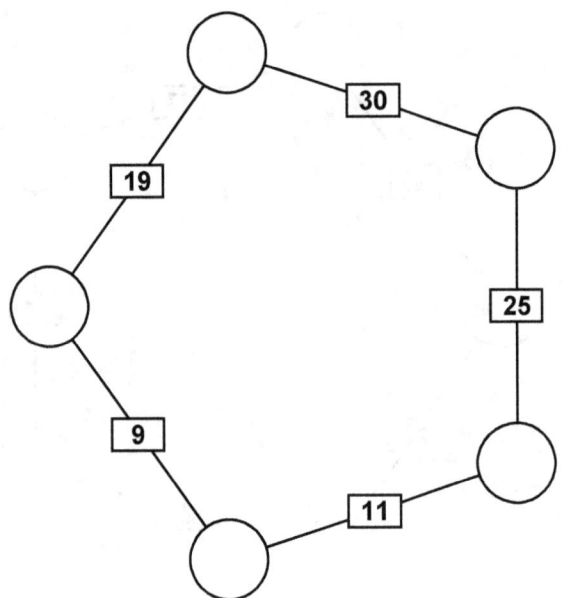

Puzzle 262

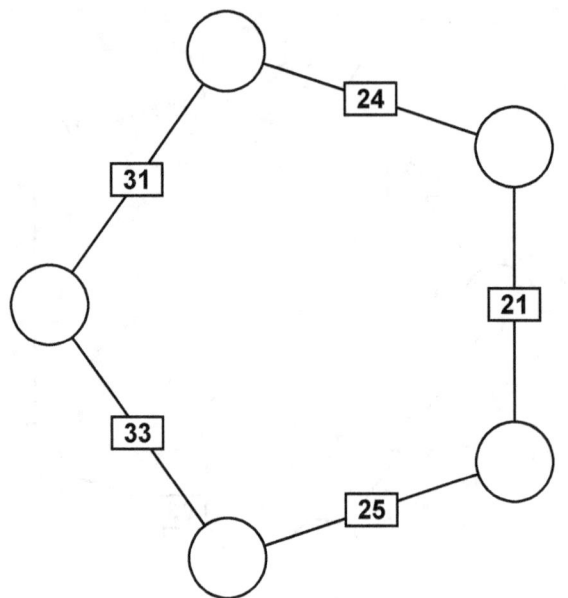

Puzzle 263

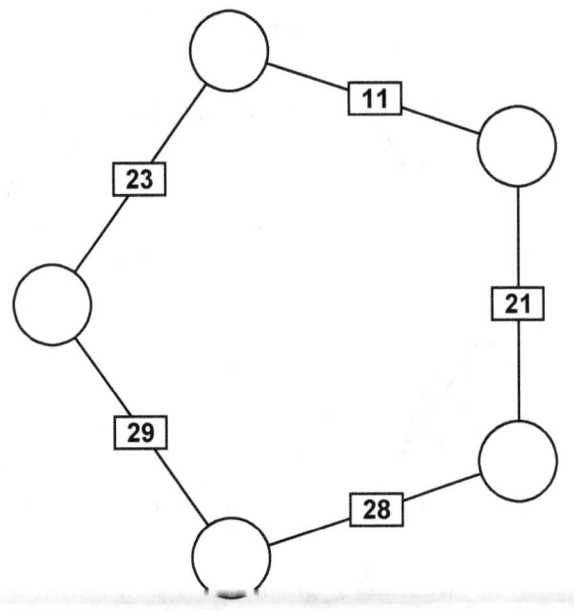

Puzzle 264

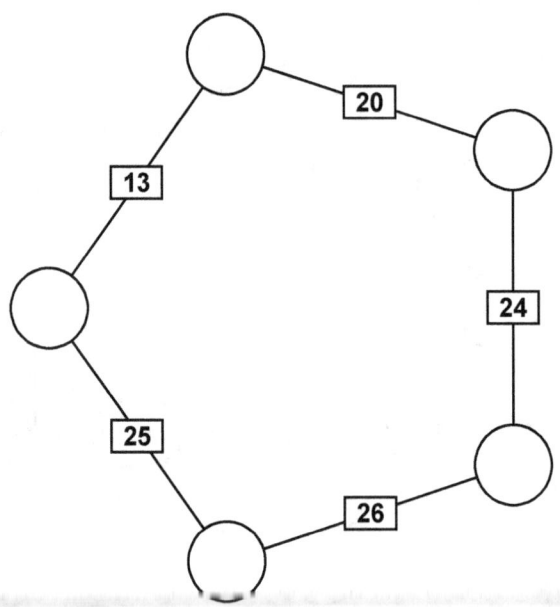

Puzzle 265

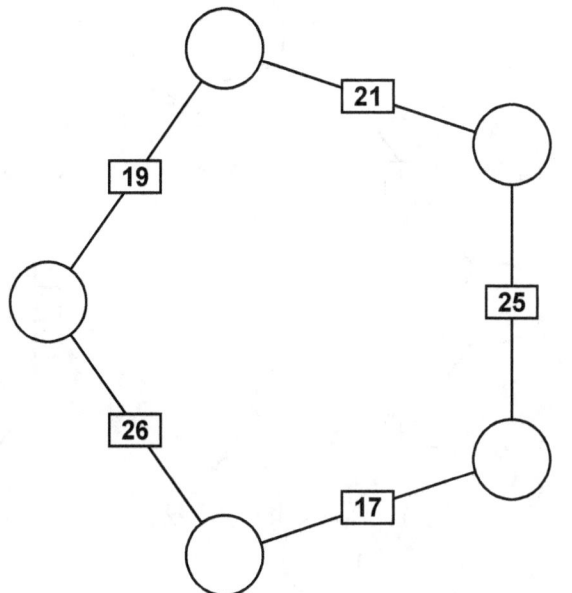

Puzzle 266

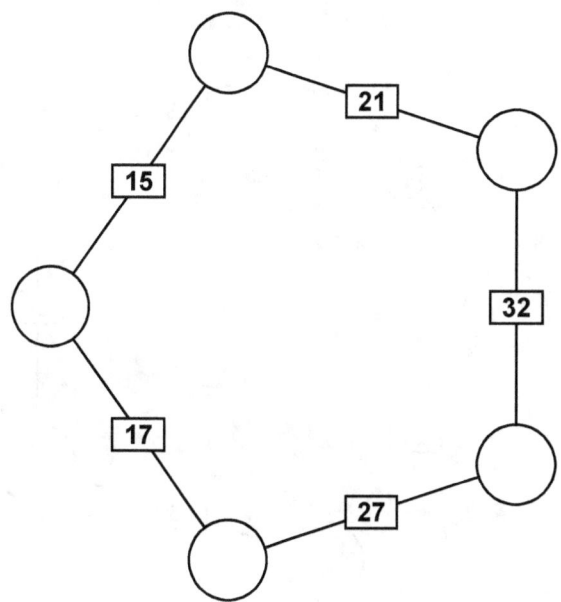

Puzzle 267

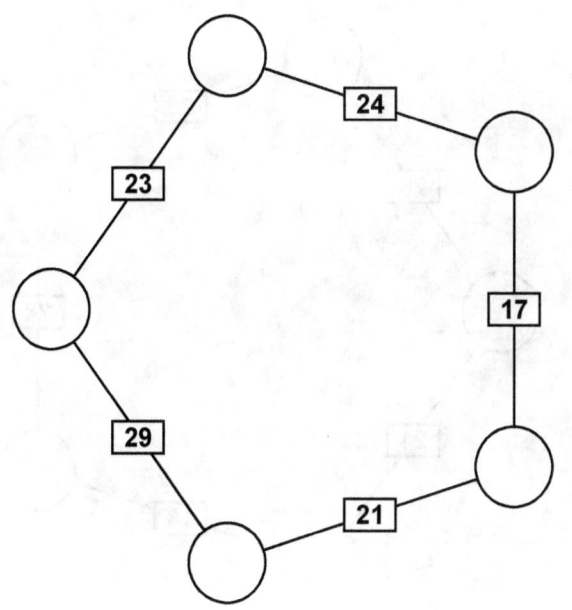

Puzzle 268

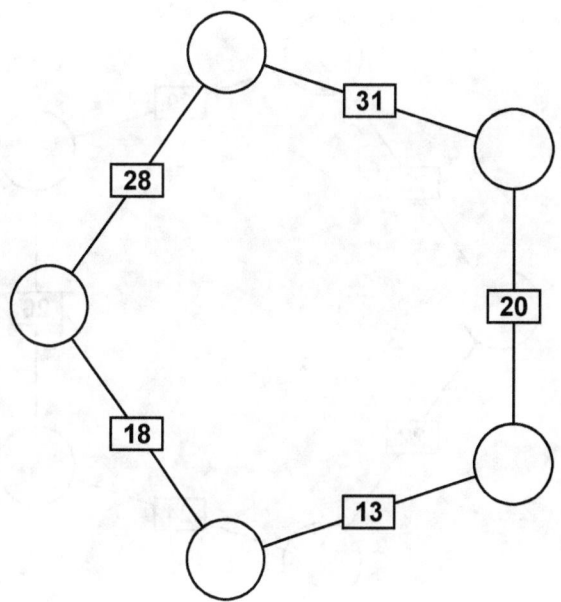

Puzzle 269

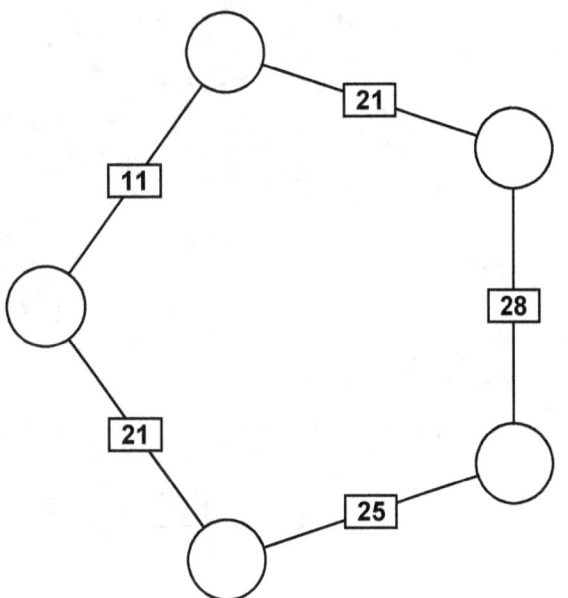

Puzzle 270

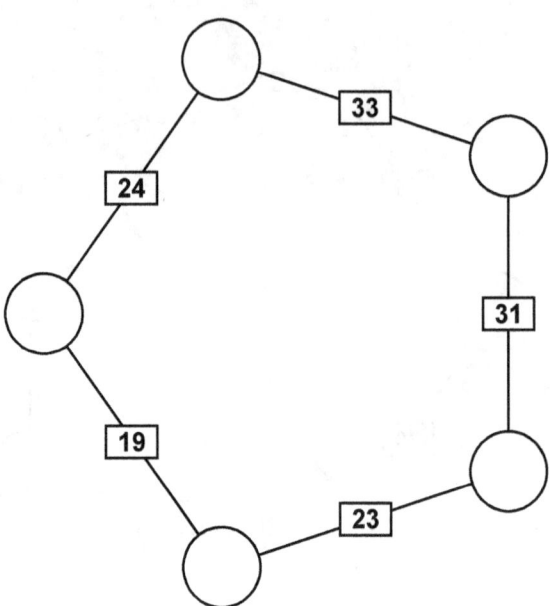

Puzzle 271

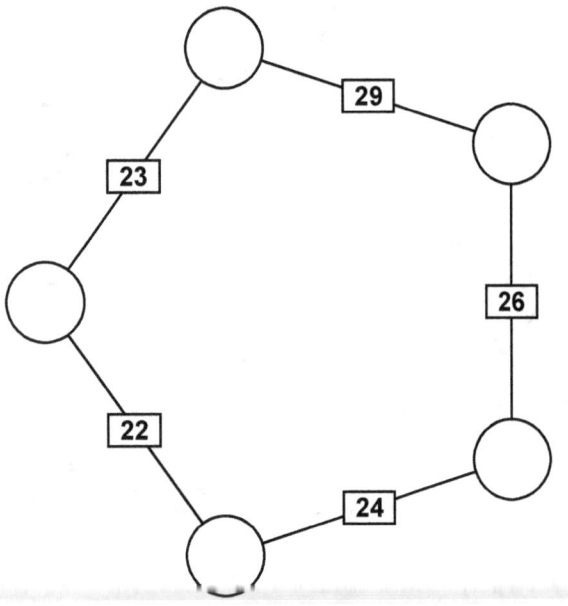

Puzzle 272

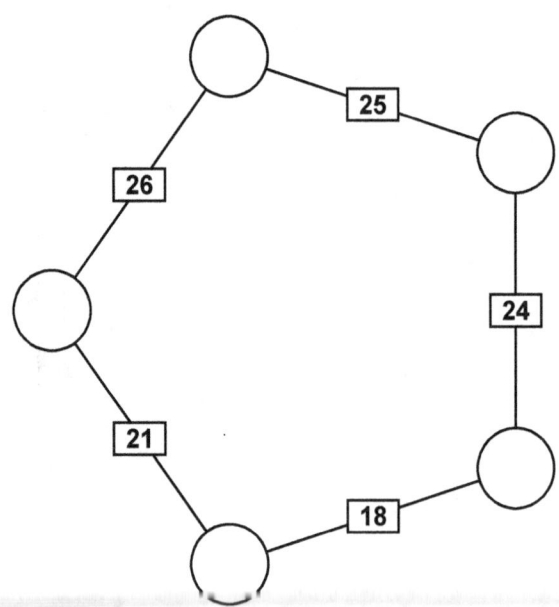

Puzzle 273

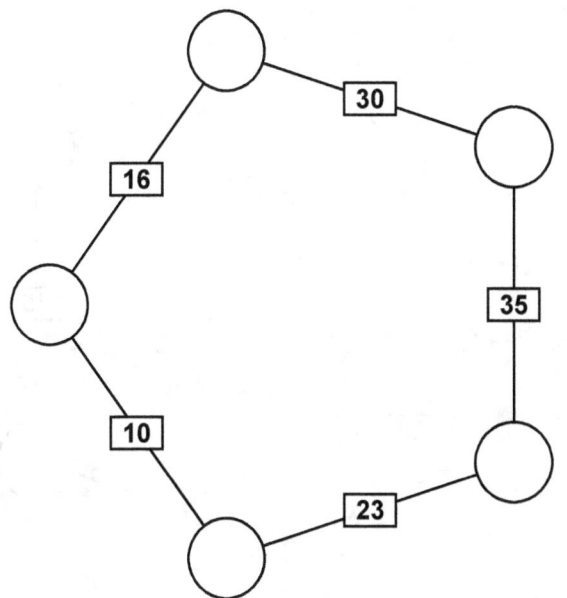

Puzzle 274

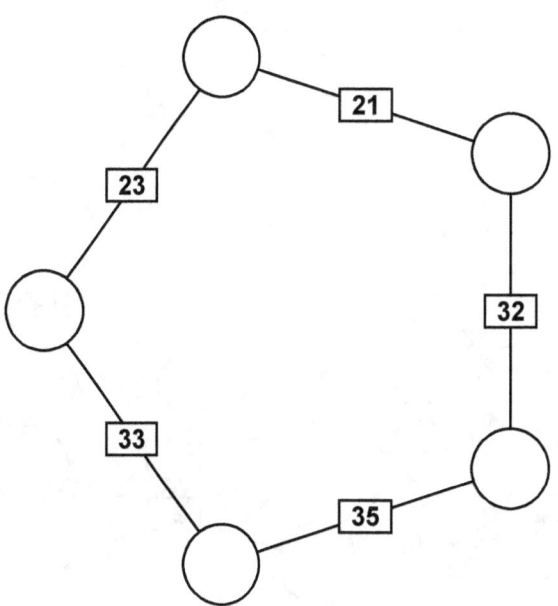

Puzzle 275

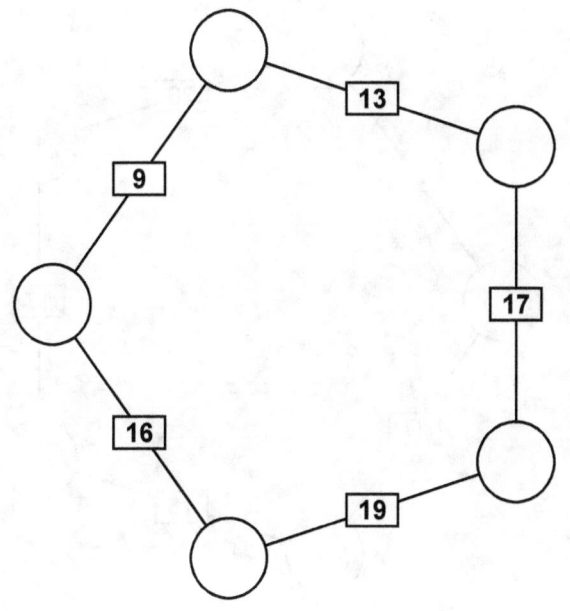

Puzzle 276

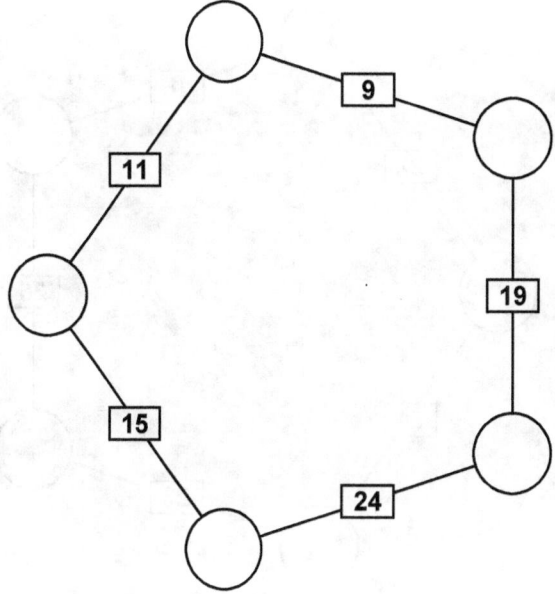

Puzzle 277

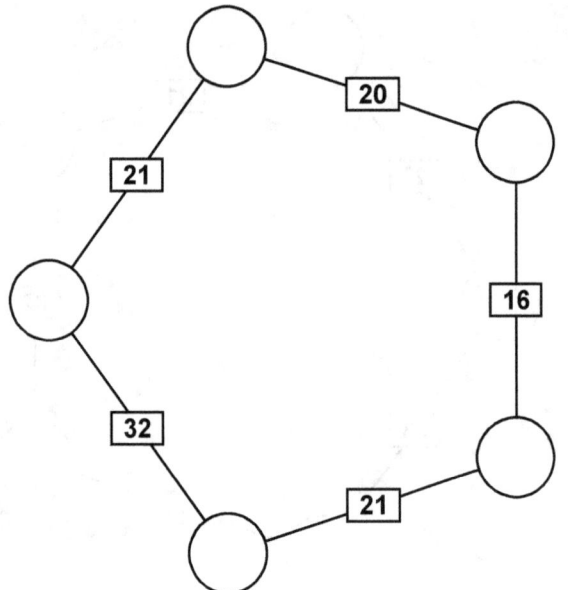

Puzzle 278

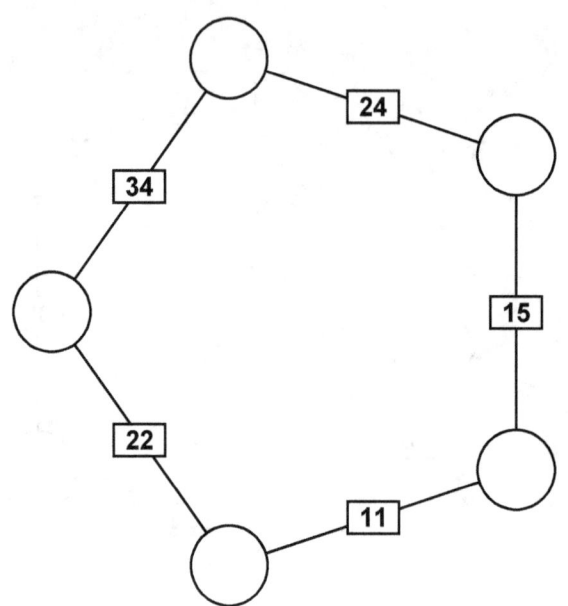

Puzzle 279

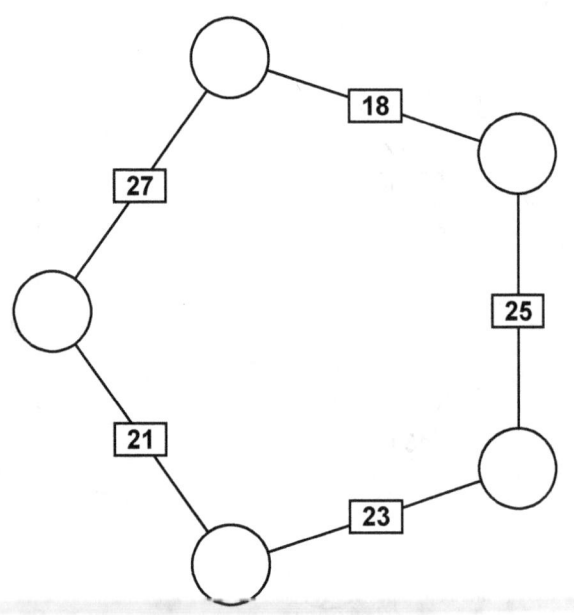

Puzzle 280

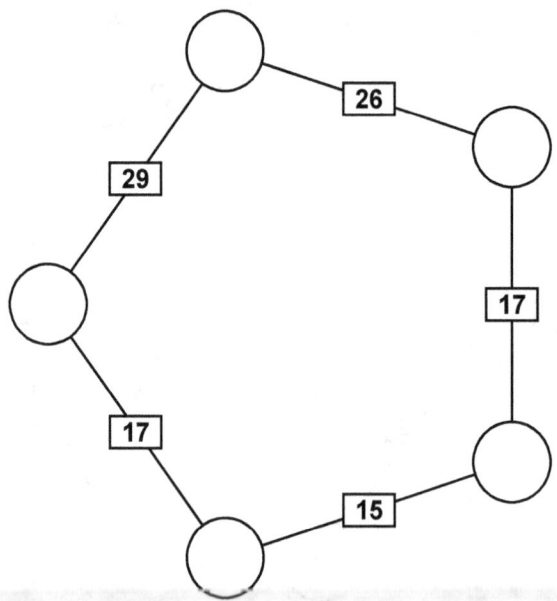

Puzzle 281

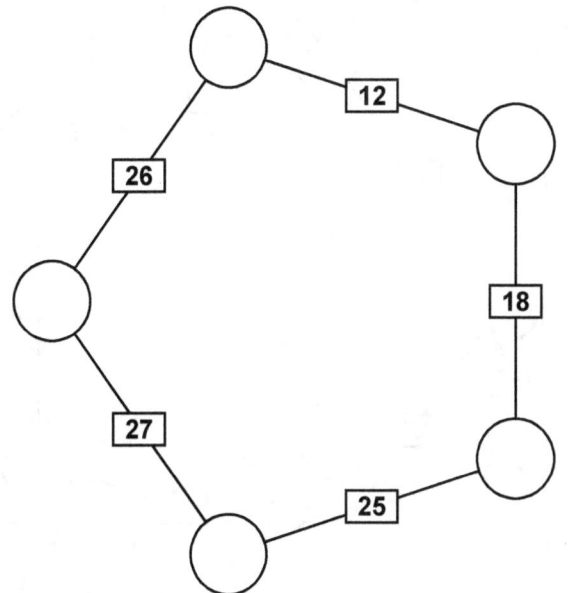

Puzzle 282

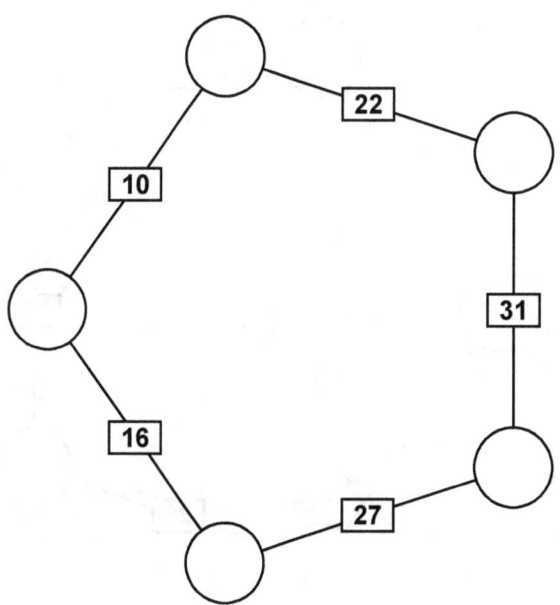

Puzzle 283

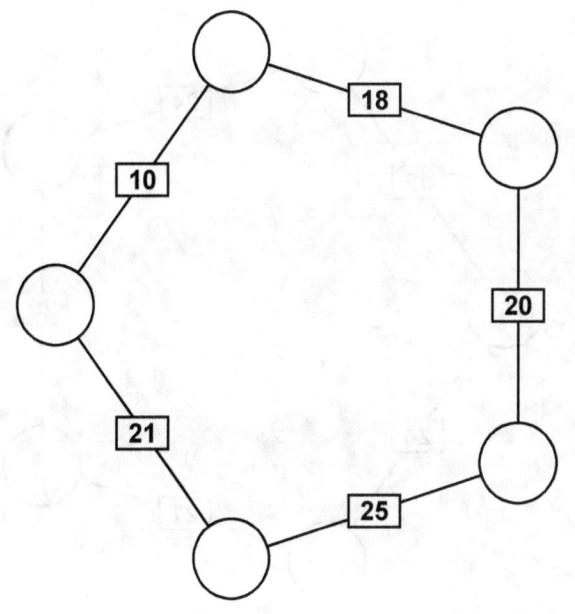

Puzzle 284

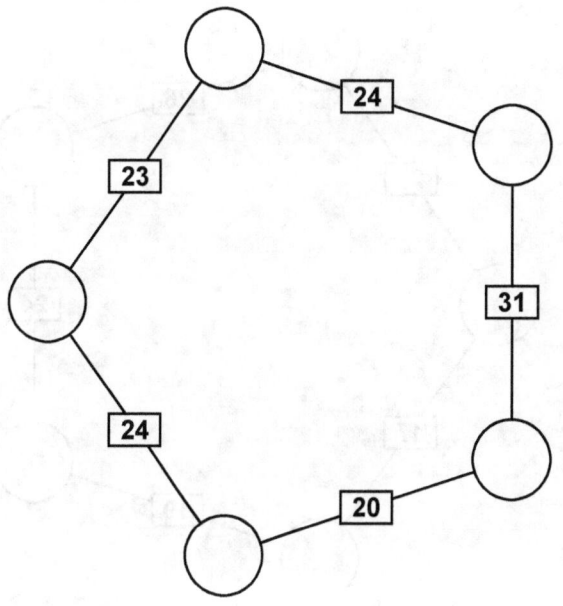

Puzzle 285

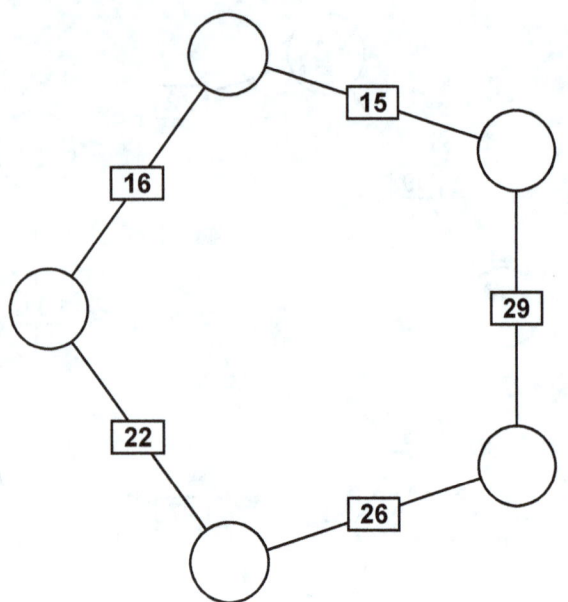

Puzzle 286

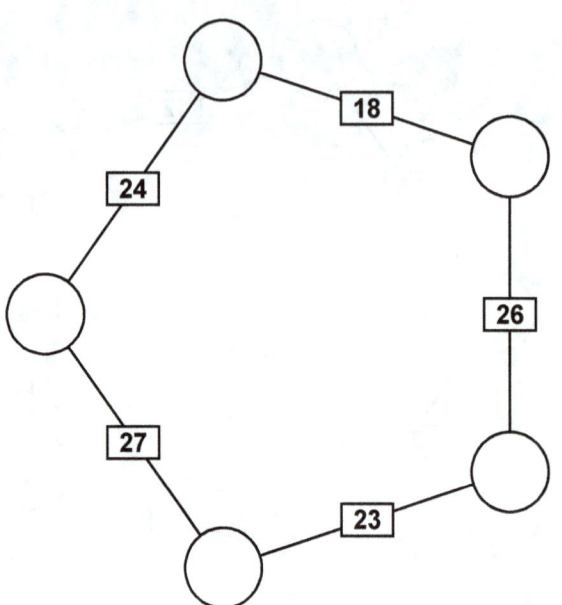

Puzzle 287

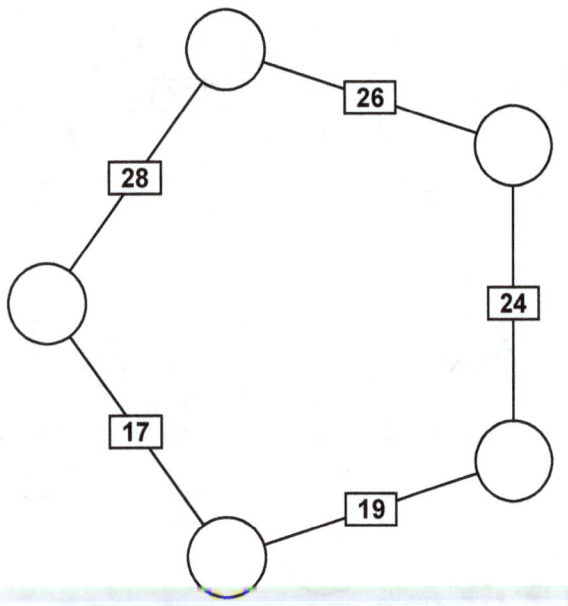

Puzzle 288

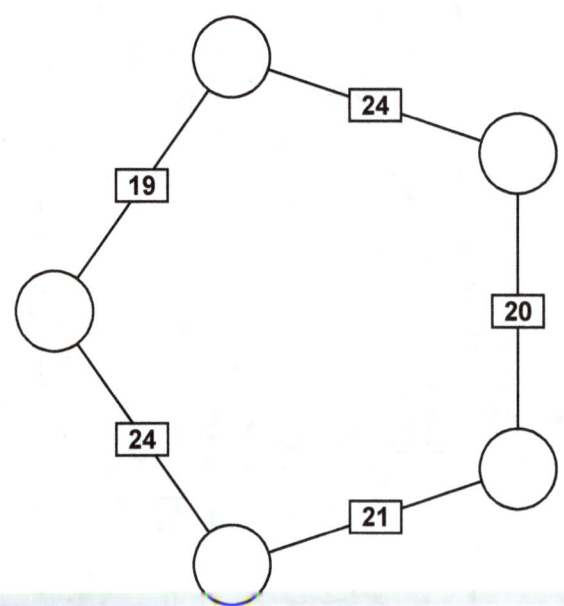

Puzzle 289

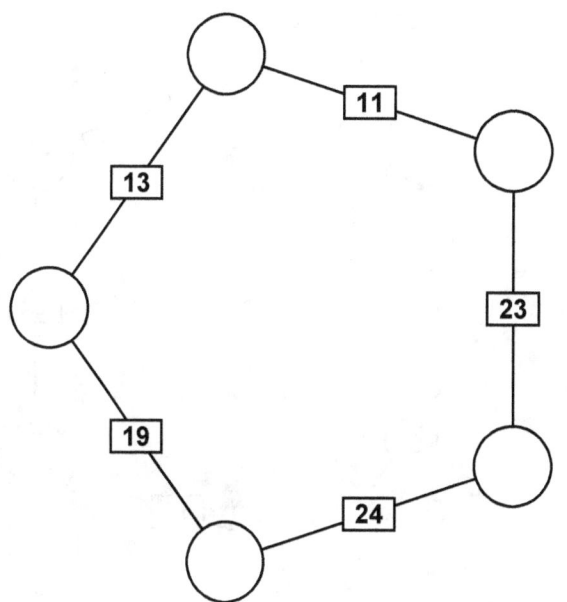

Puzzle 290

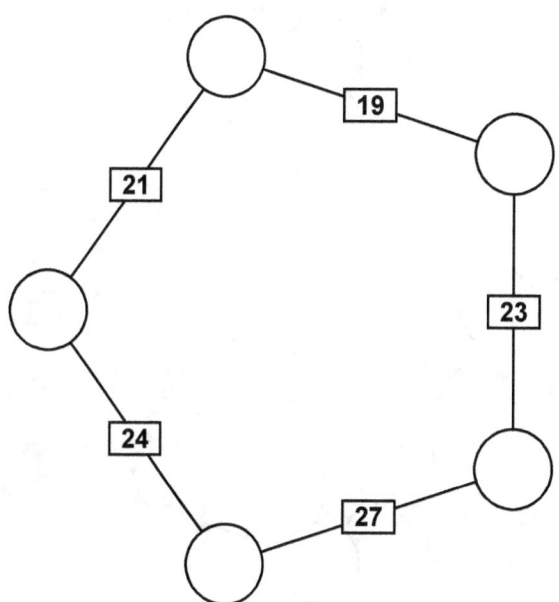

Puzzle 291

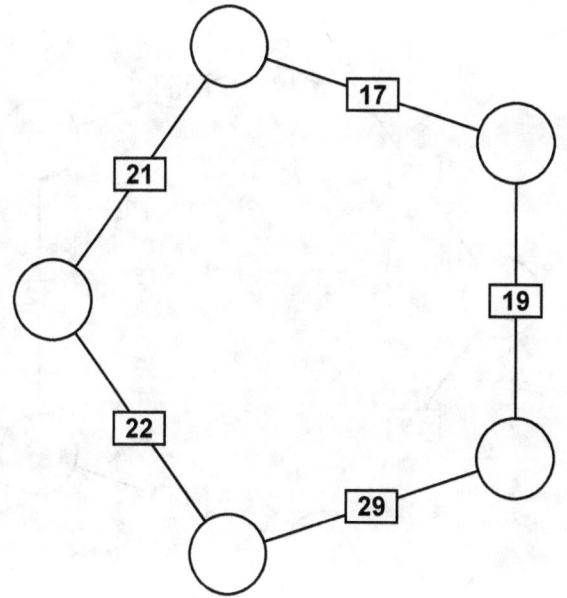

Puzzle 292

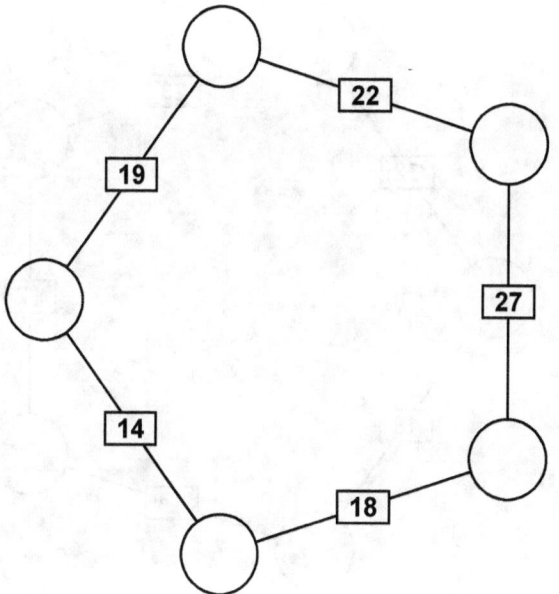

Puzzle 293

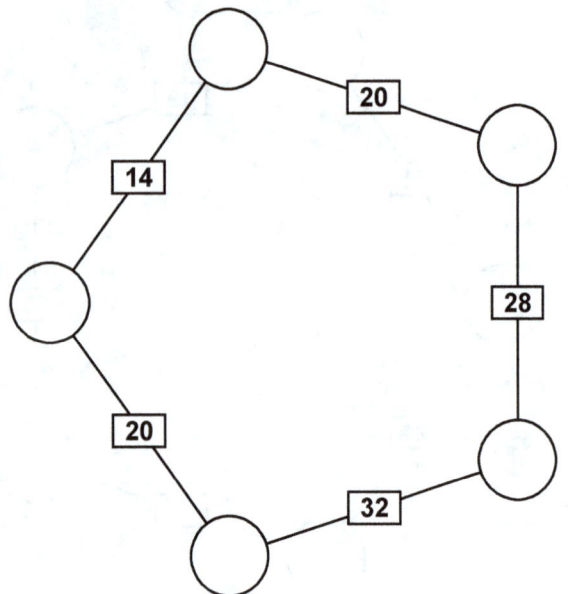

Puzzle 294

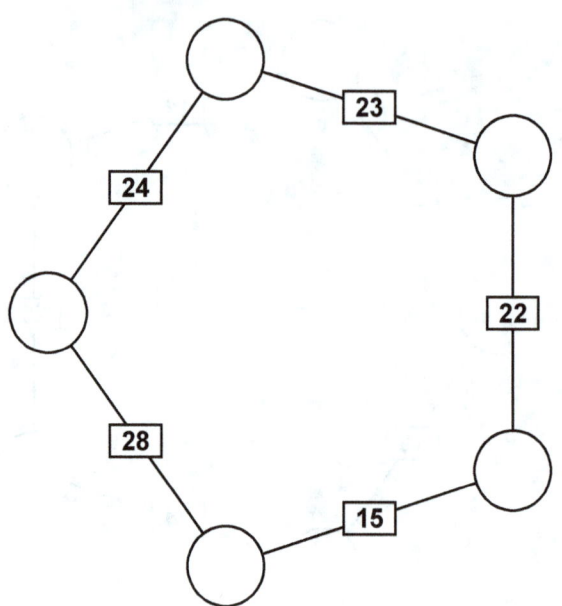

Puzzle 295

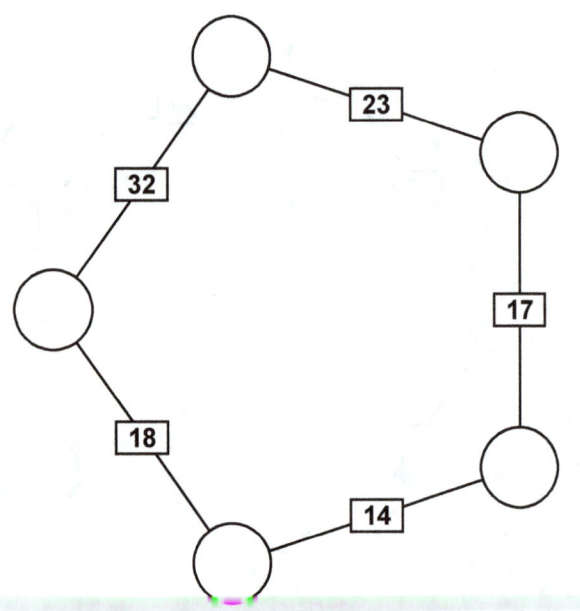

Puzzle 296

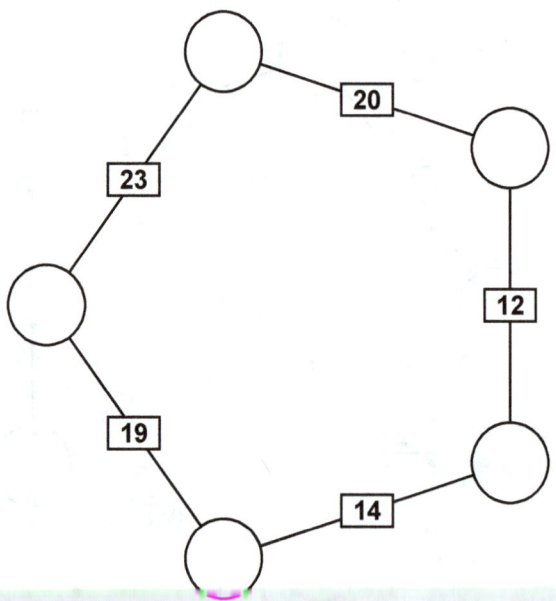

Puzzle 297

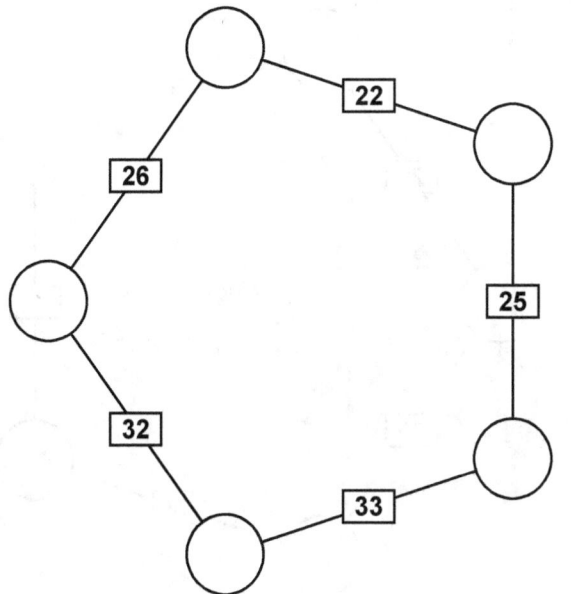

Puzzle 298

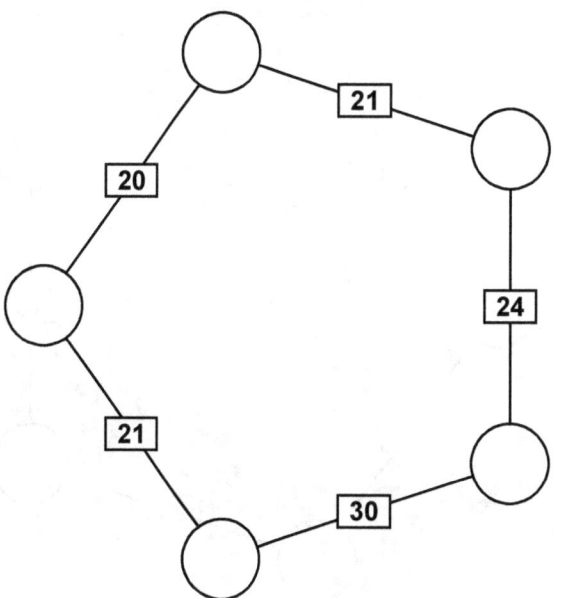

Puzzle 299

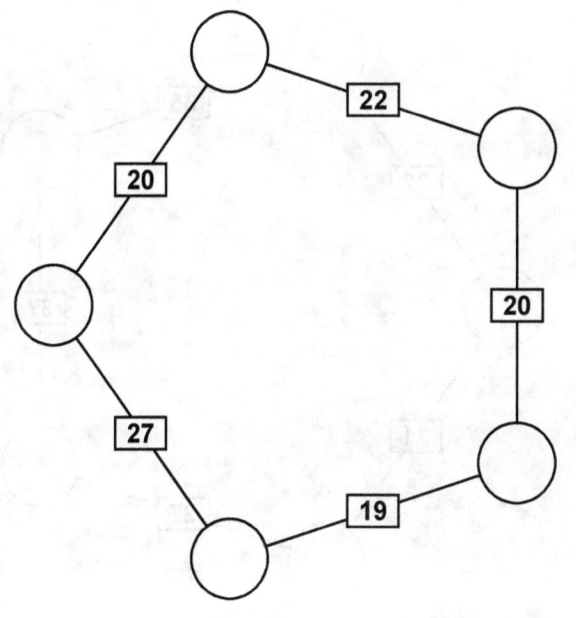

Puzzle 300

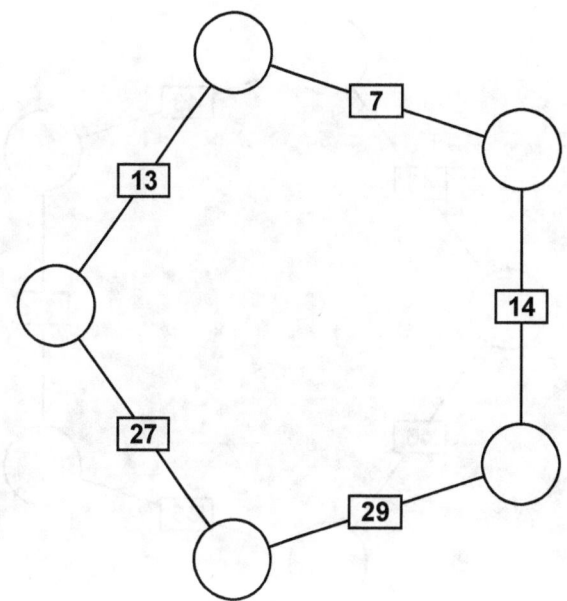

Puzzle 301

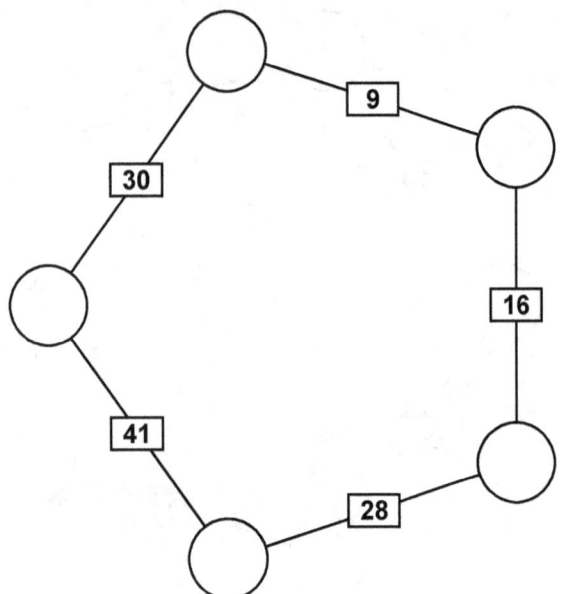

Puzzle 302

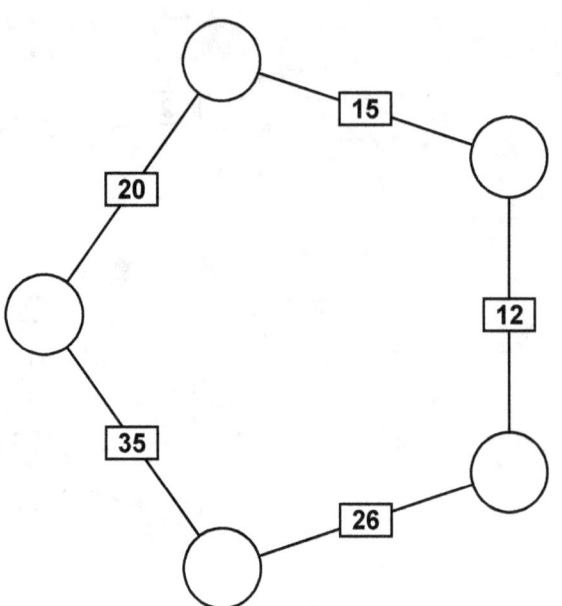

Puzzle 303

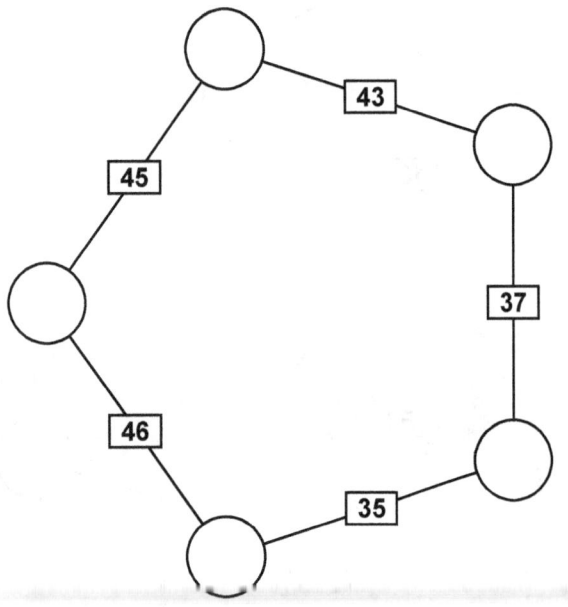

Puzzle 304

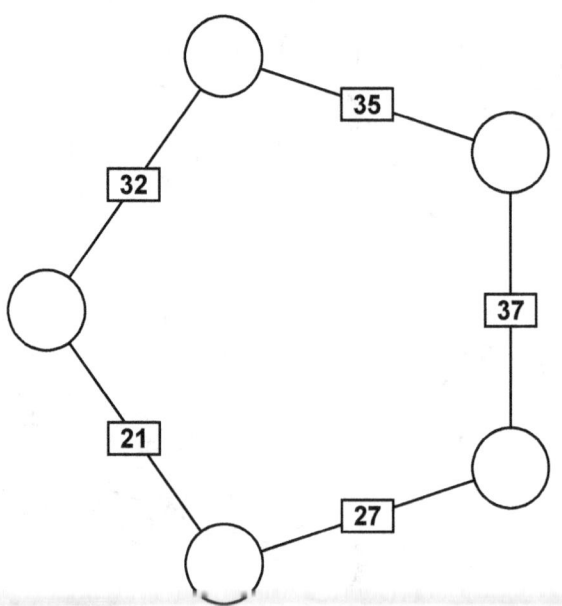

Puzzle 305

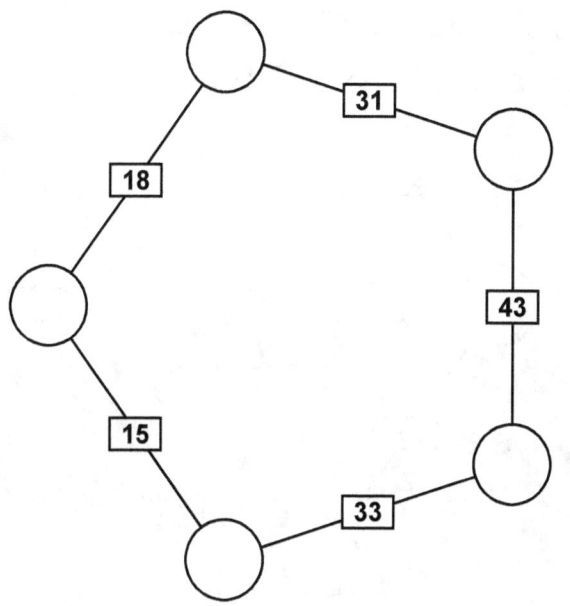

Puzzle 306

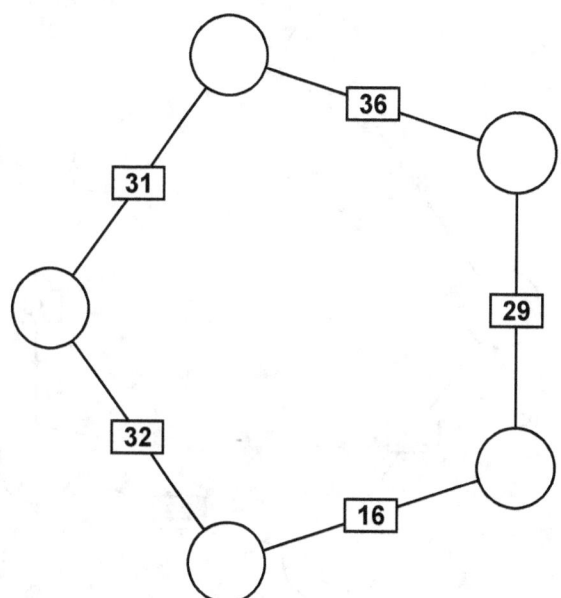

Puzzle 307

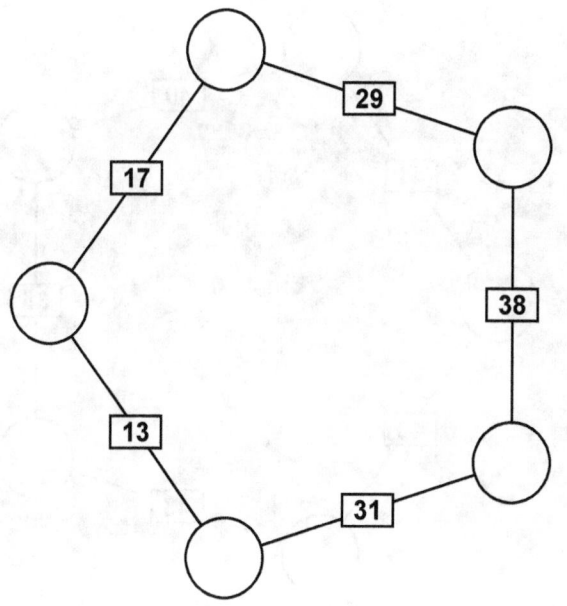

Puzzle 308

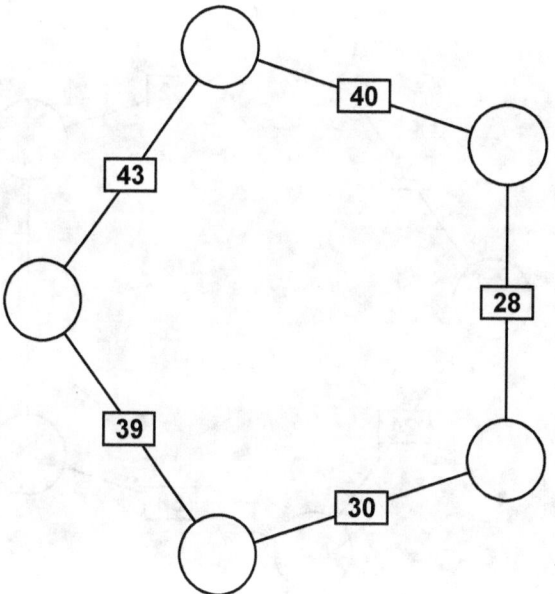

Puzzle 309

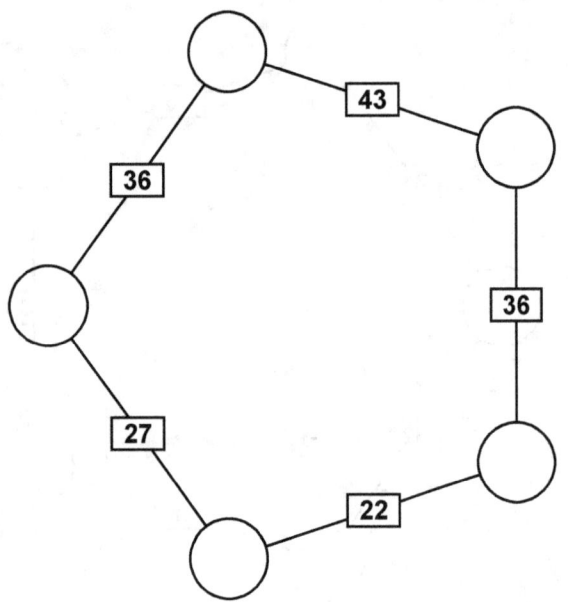

Puzzle 310

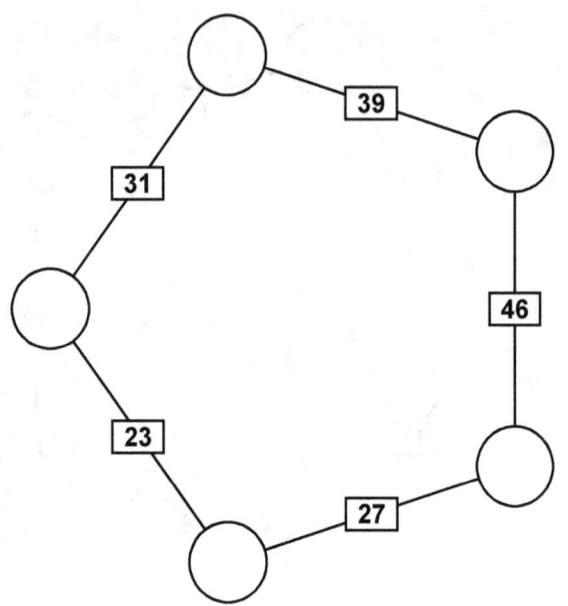

Puzzle 311

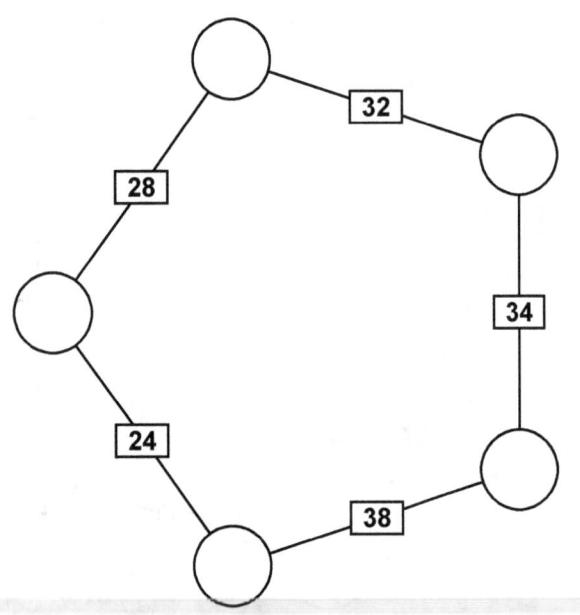

Puzzle 312

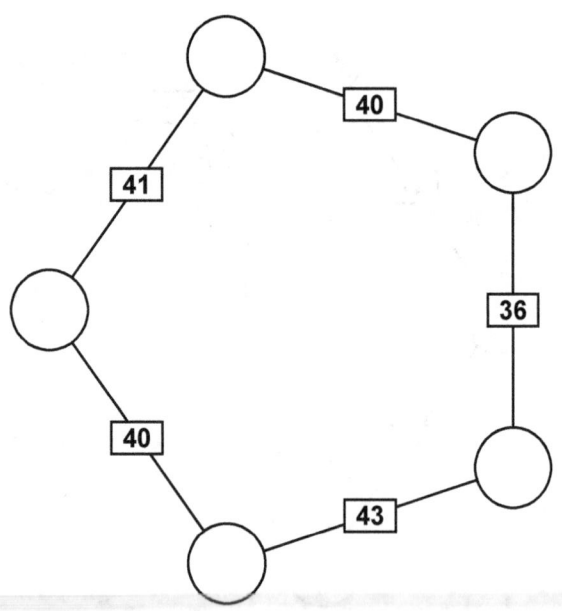

Puzzle 313

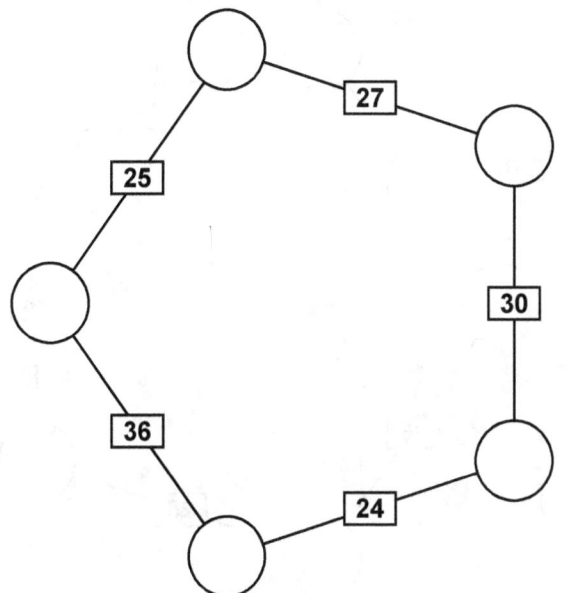

Puzzle 314

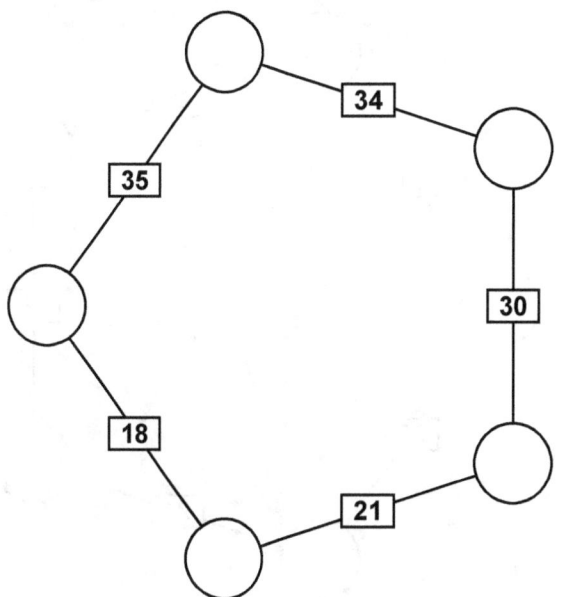

Puzzle 315

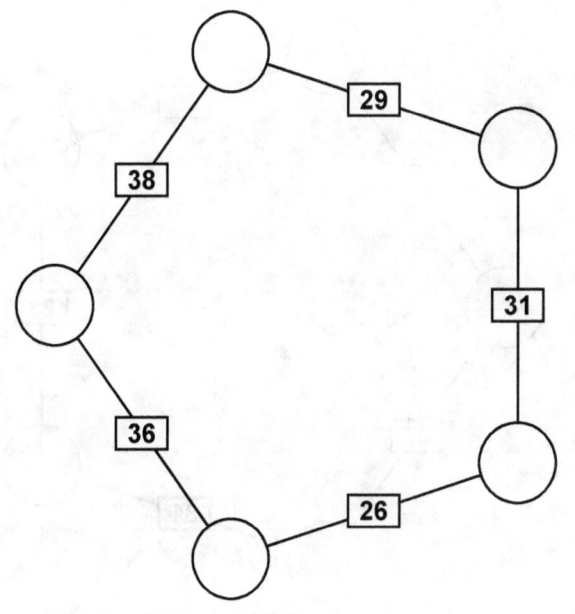

Puzzle 316

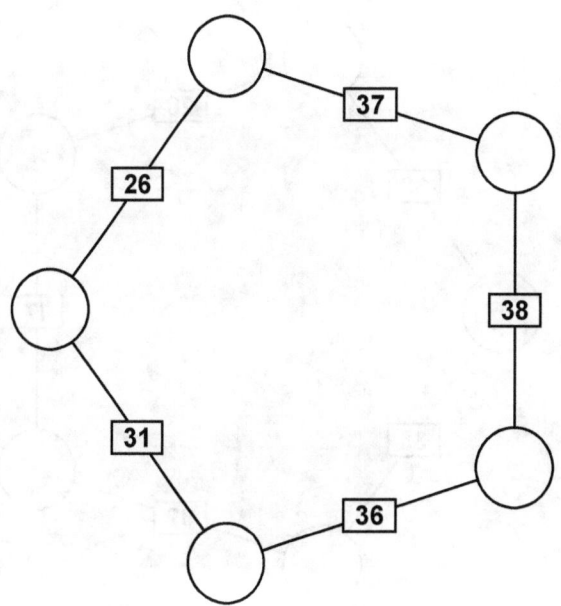

Puzzle 317

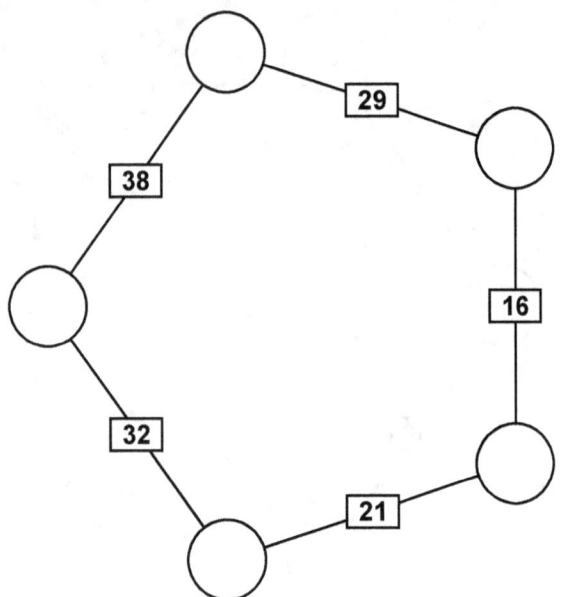

Puzzle 318

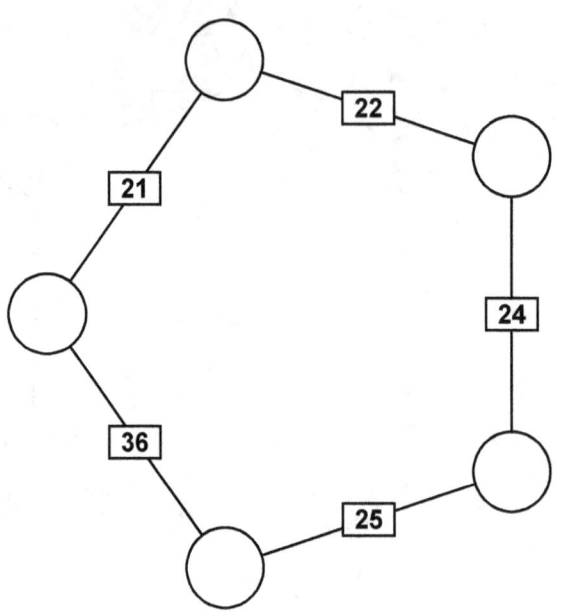

Puzzle 319

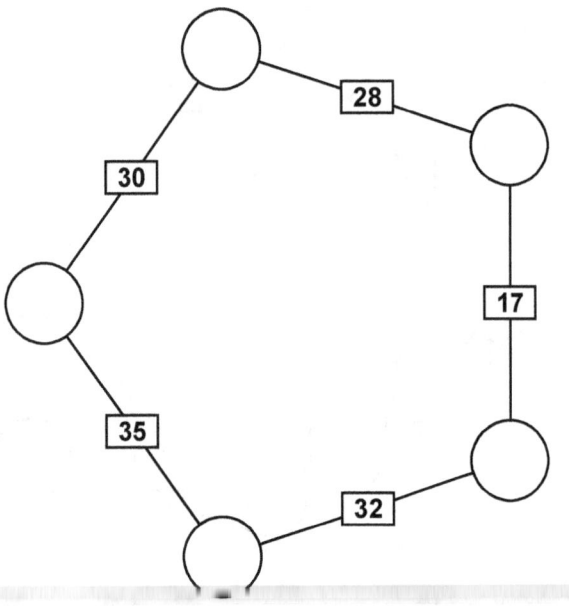

Puzzle 320

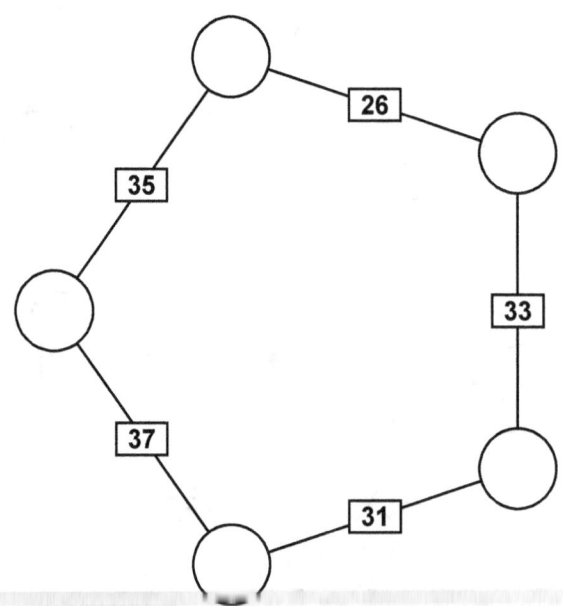

Puzzle 321

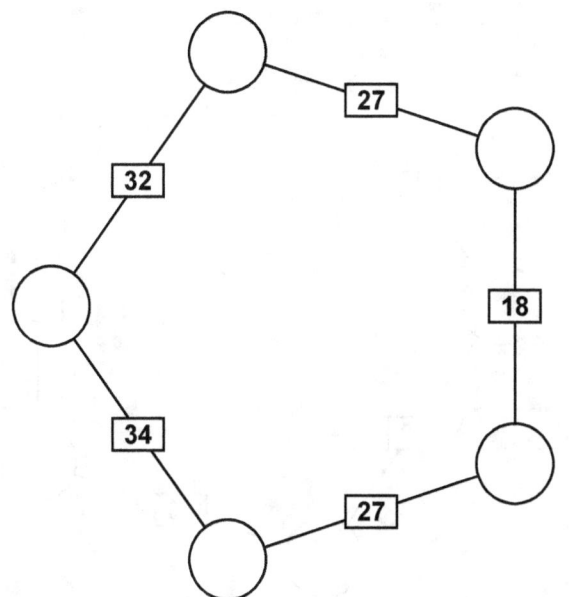

Puzzle 322

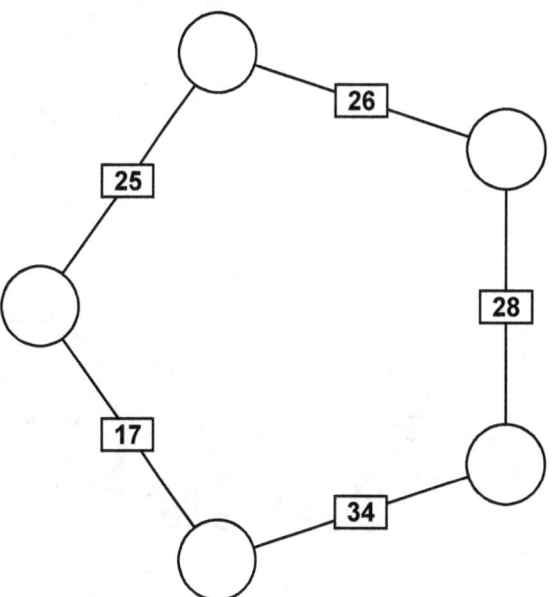

Puzzle 323

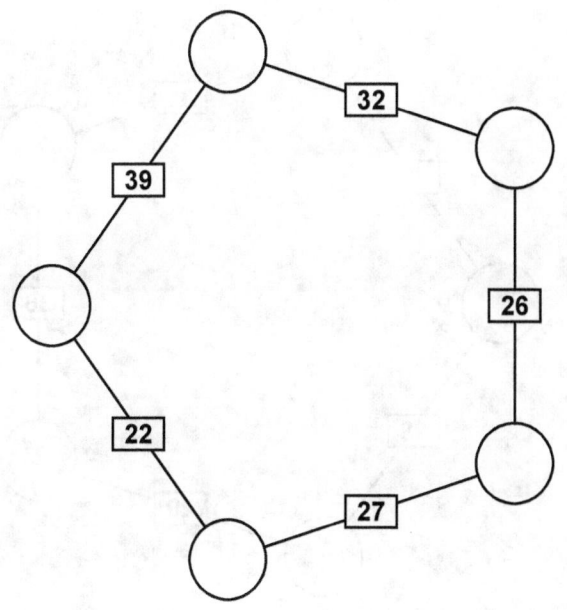

Puzzle 324

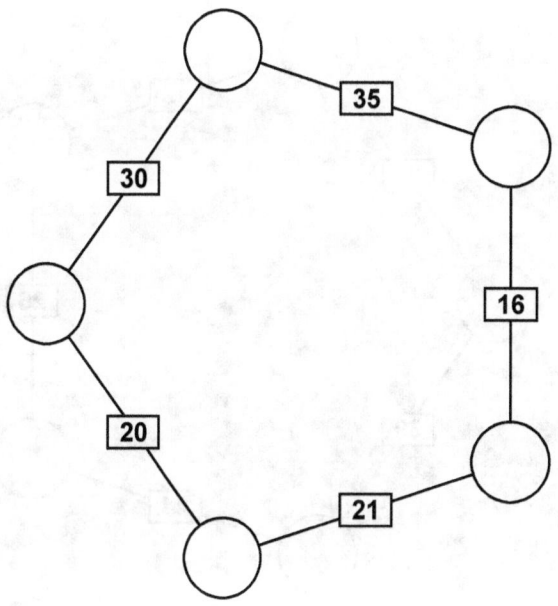

Puzzle 325

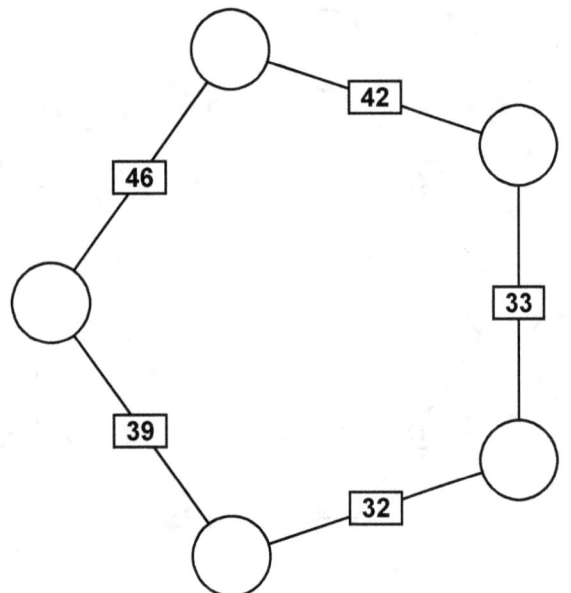

Puzzle 326

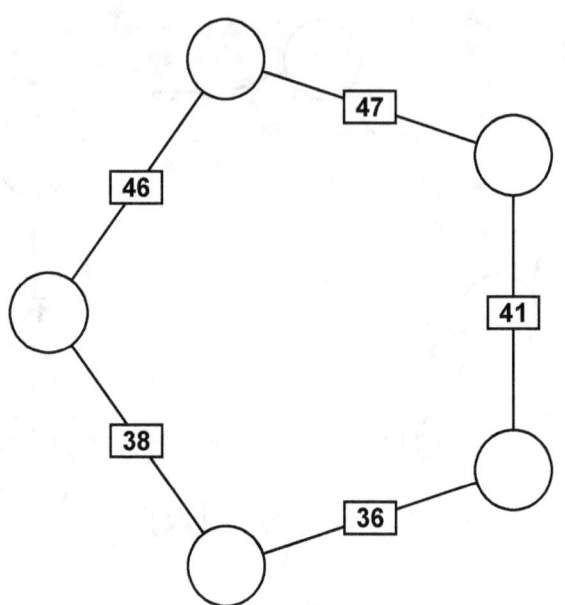

Puzzle 327

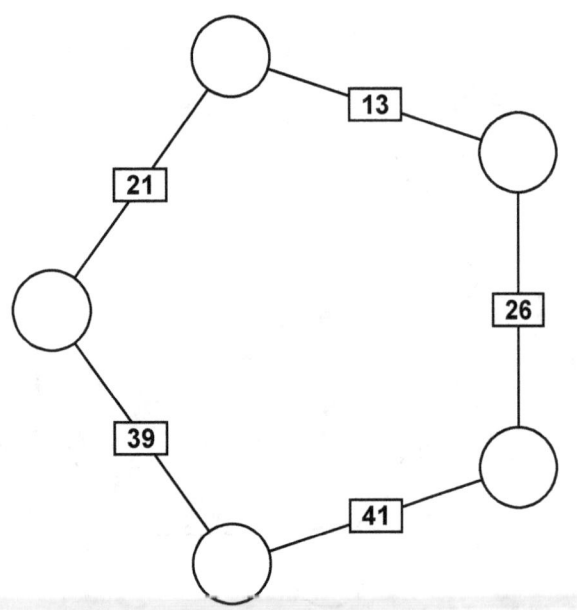

Puzzle 328

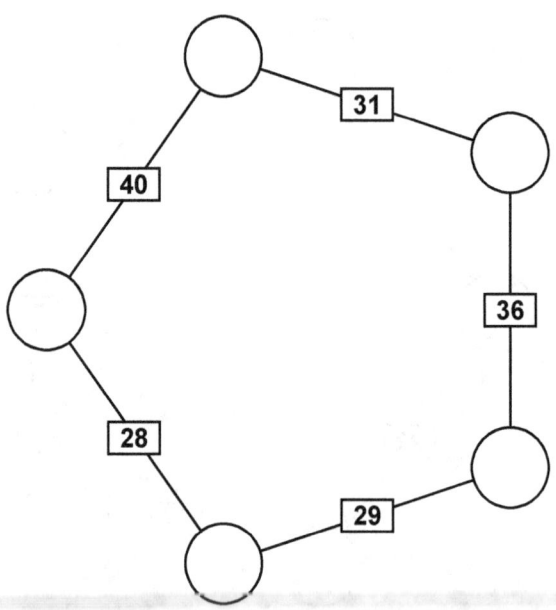

Puzzle 329

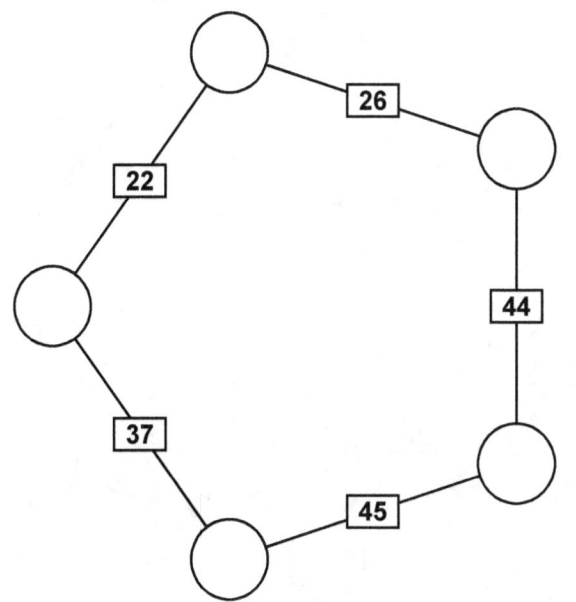

Puzzle 330

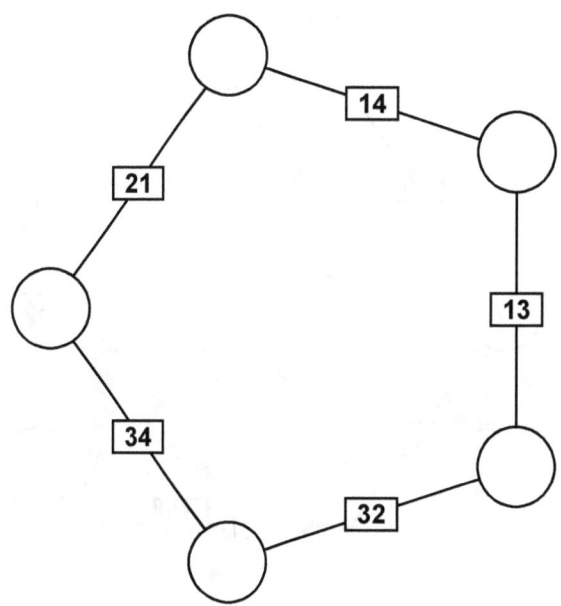

Puzzle 331

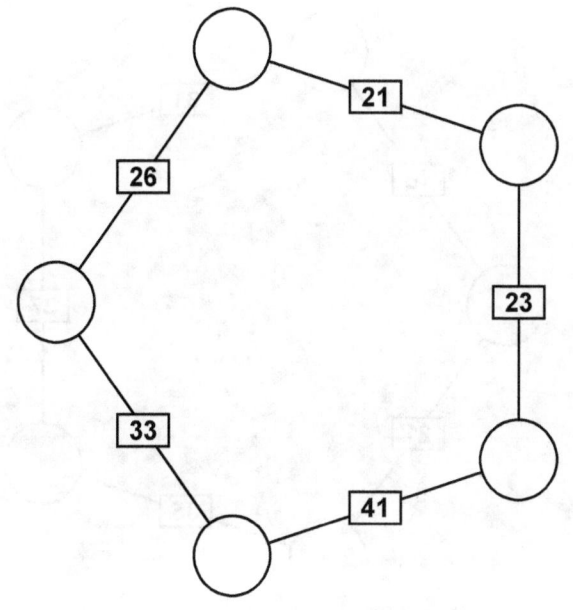

Puzzle 332

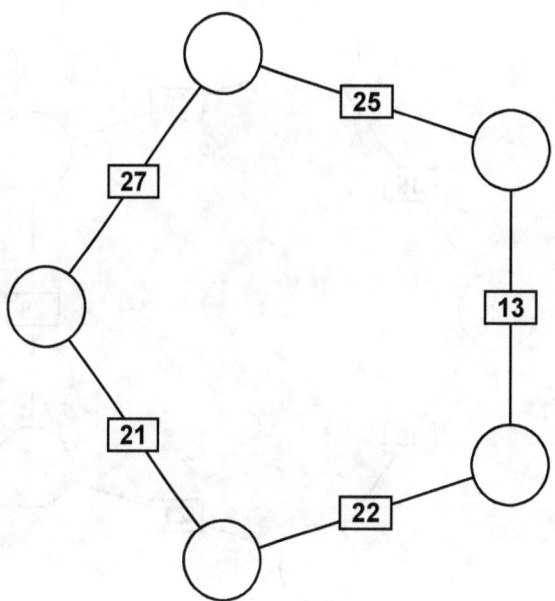

Puzzle 333

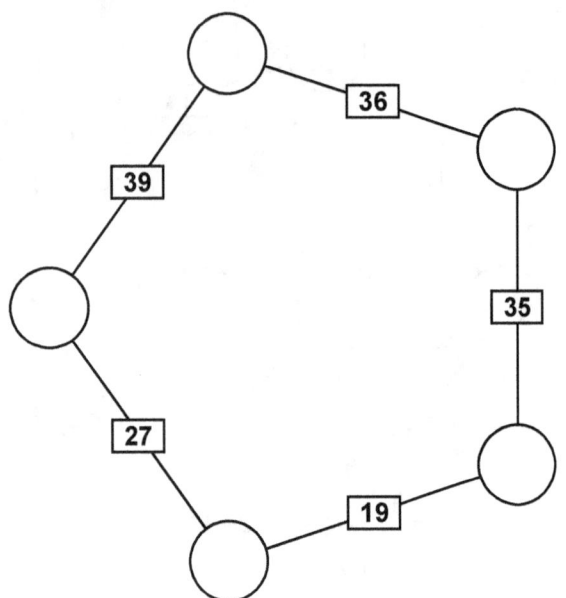

Puzzle 334

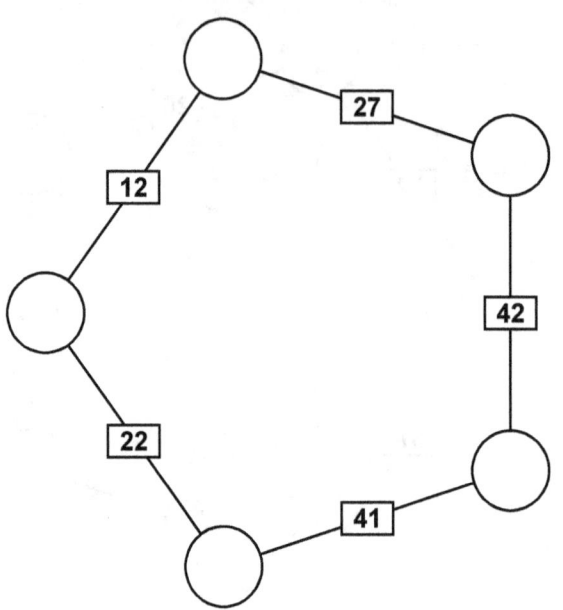

Puzzle 335

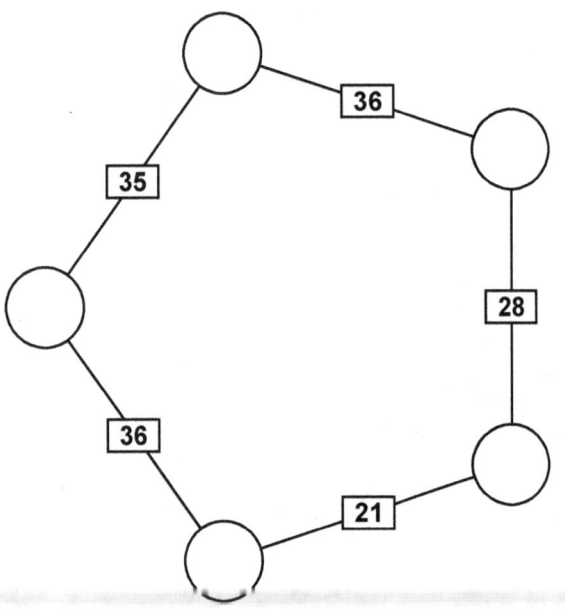

Puzzle 336

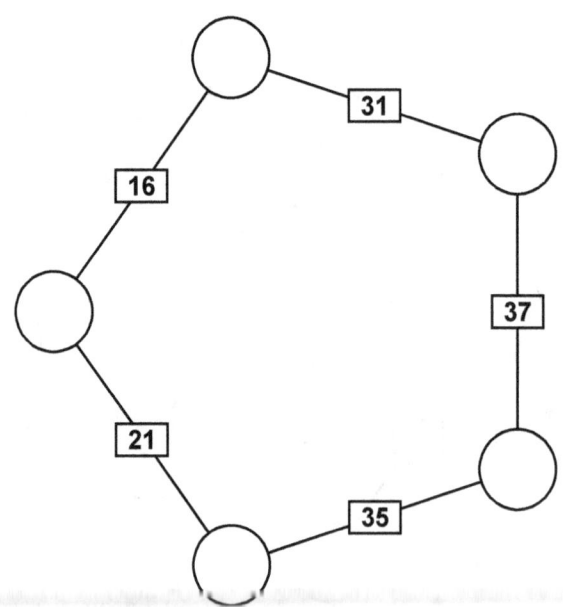

Puzzle 337

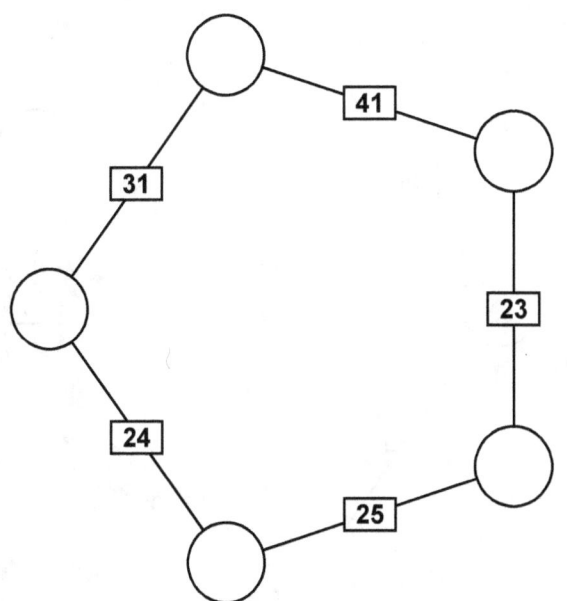

Puzzle 338

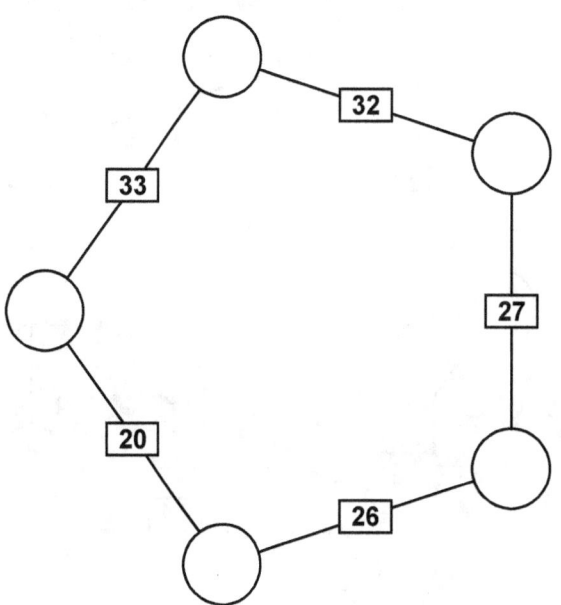

Puzzle 339

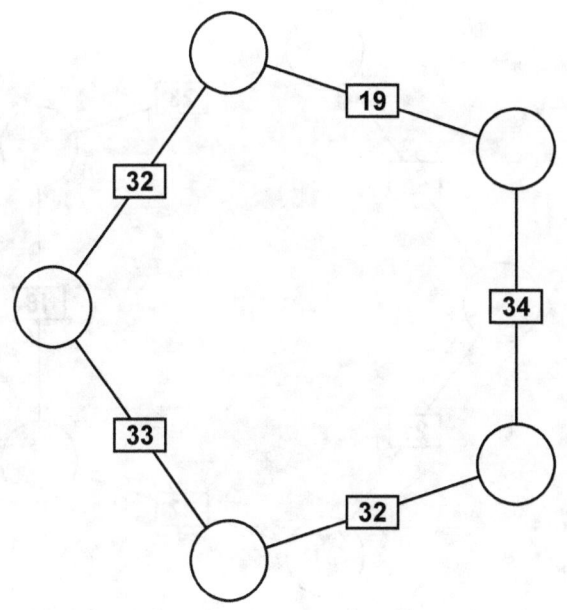

Puzzle 340

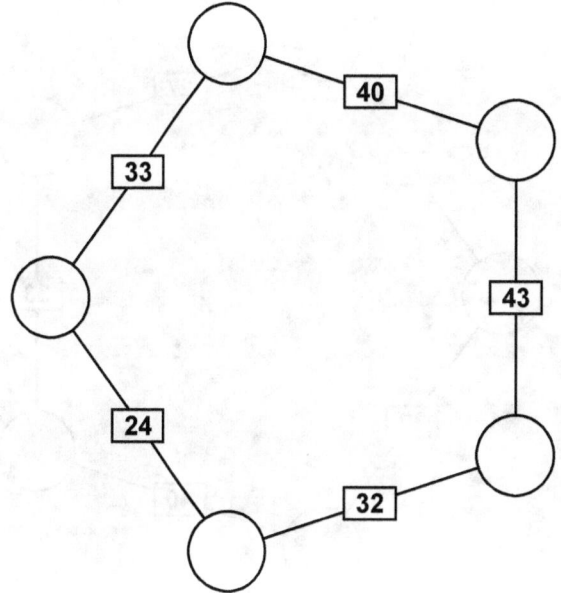

Puzzle 341

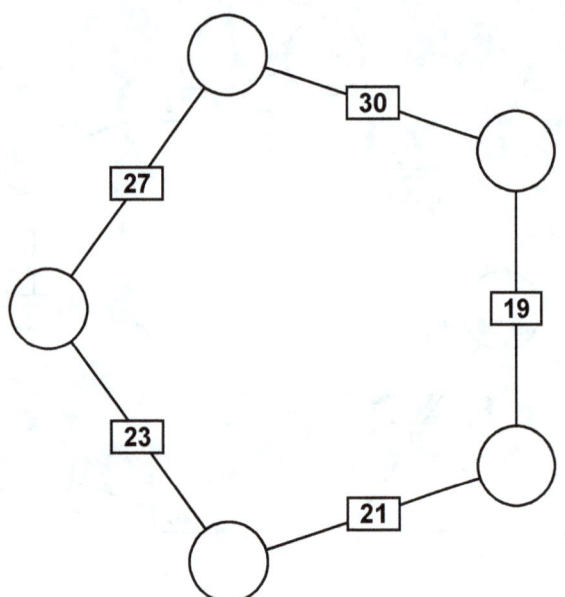

Puzzle 342

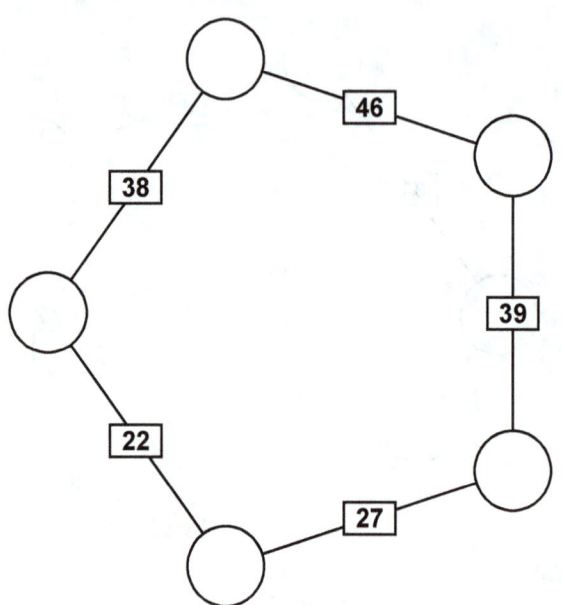

Puzzle 343

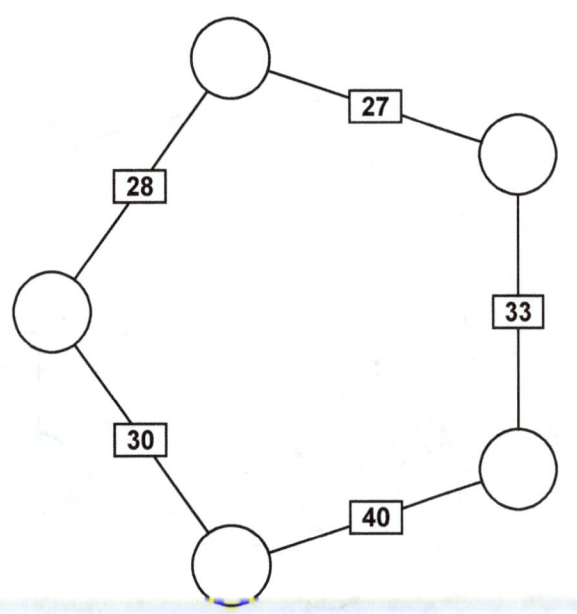

Puzzle 344

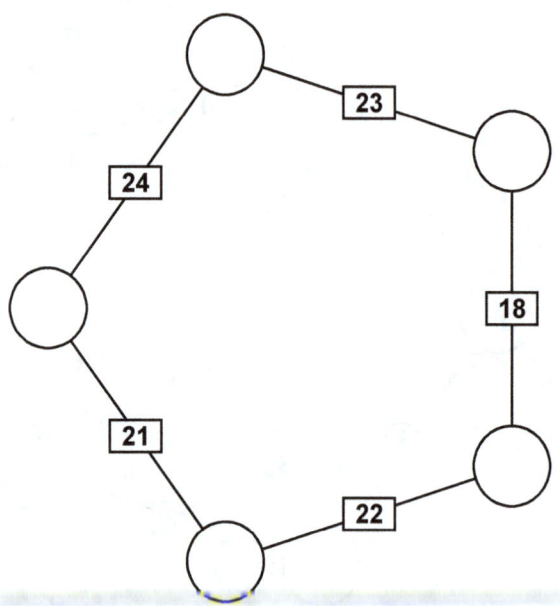

Puzzle 345

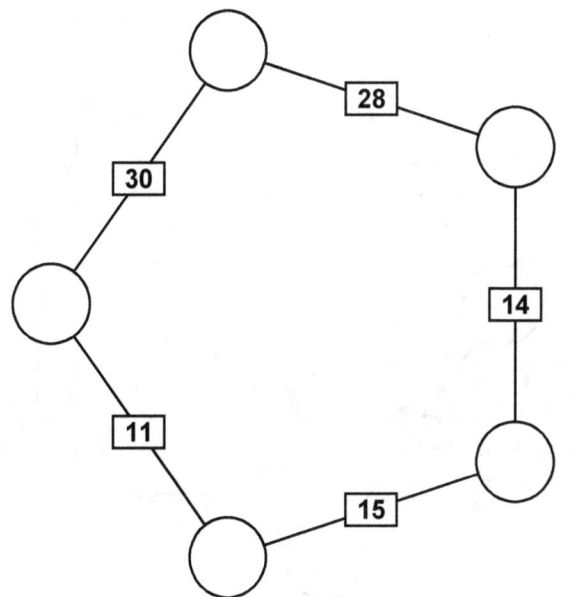

Puzzle 346

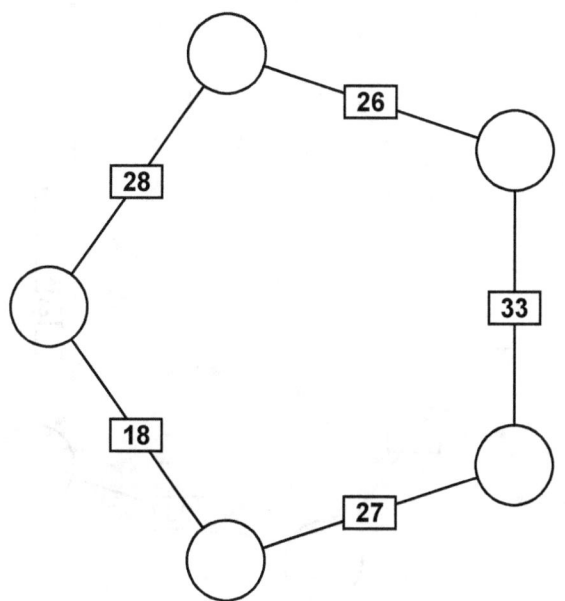

Puzzle 347

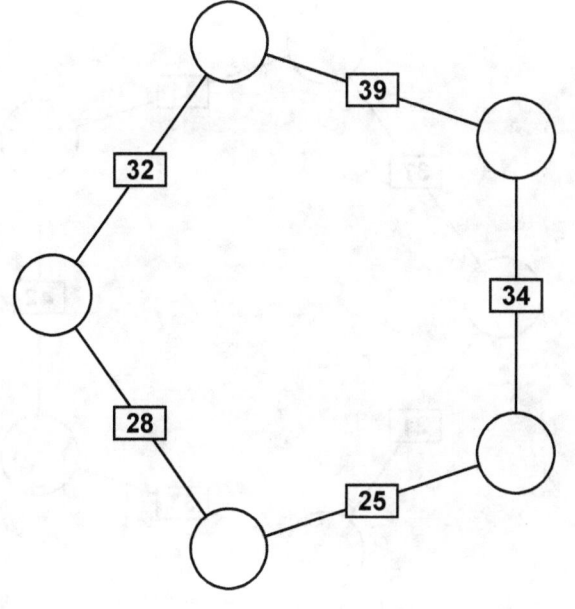

Puzzle 348

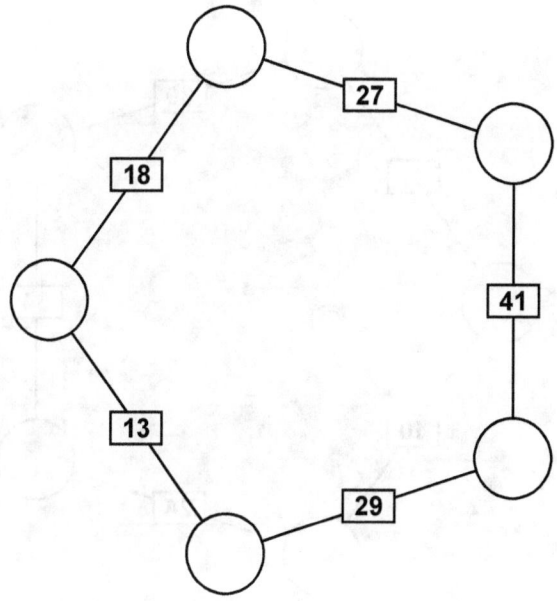

Puzzle 349

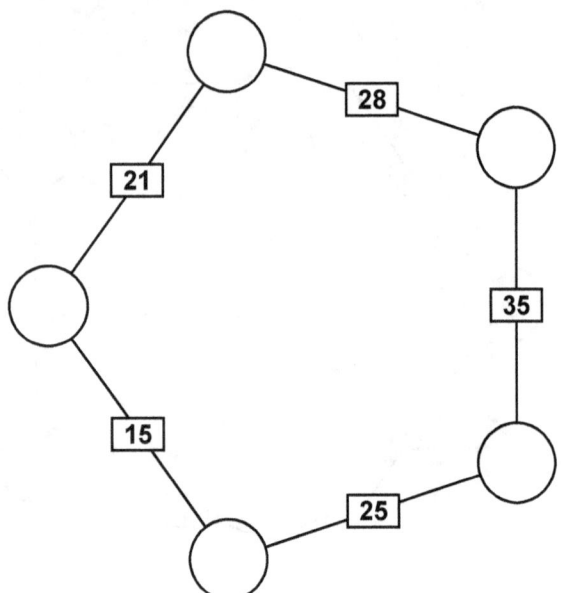

Puzzle 350

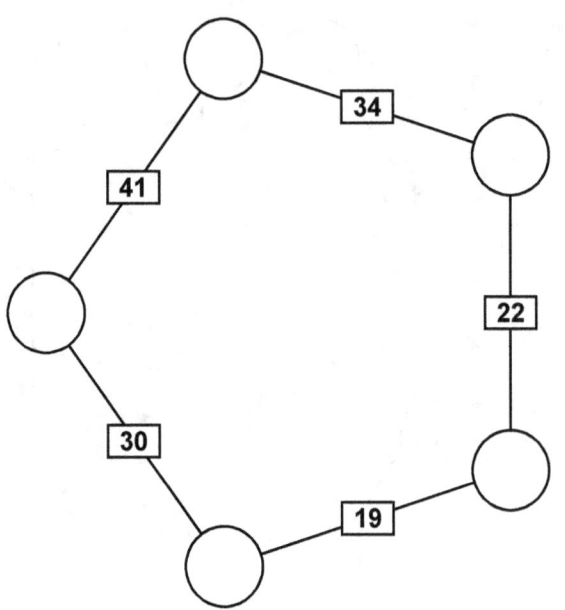

Puzzle 351

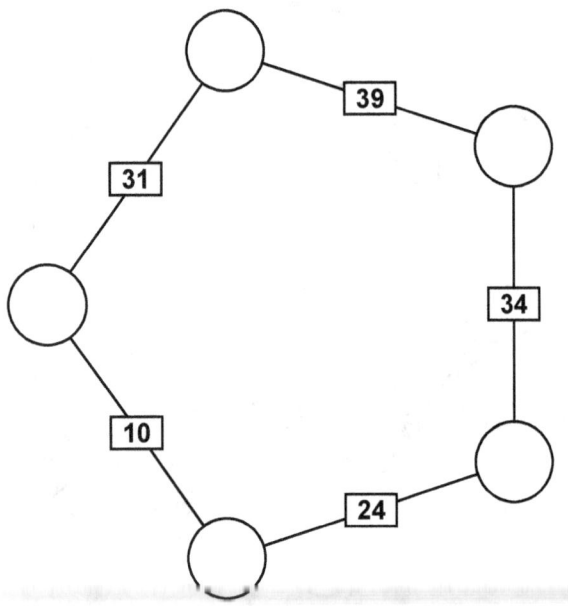

Puzzle 352

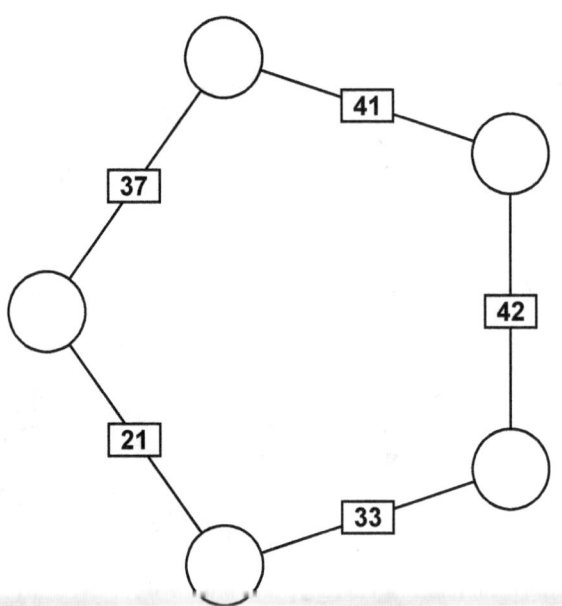

Puzzle 353

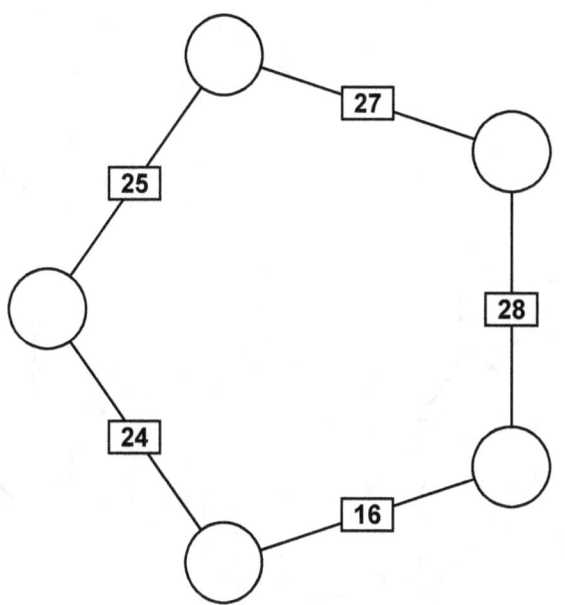

Puzzle 354

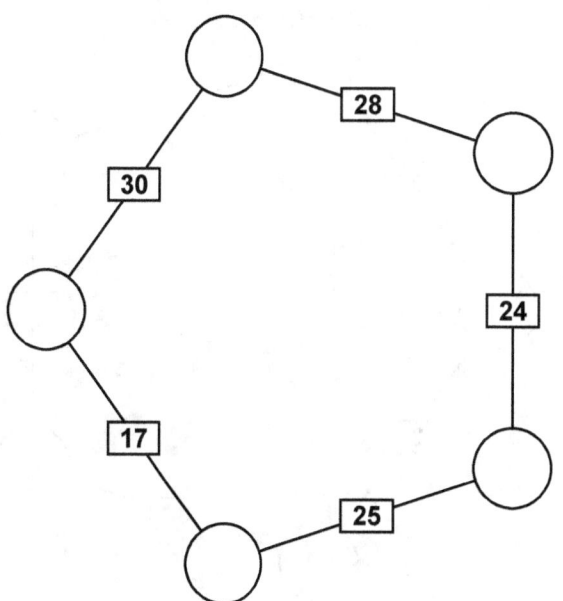

Puzzle 355

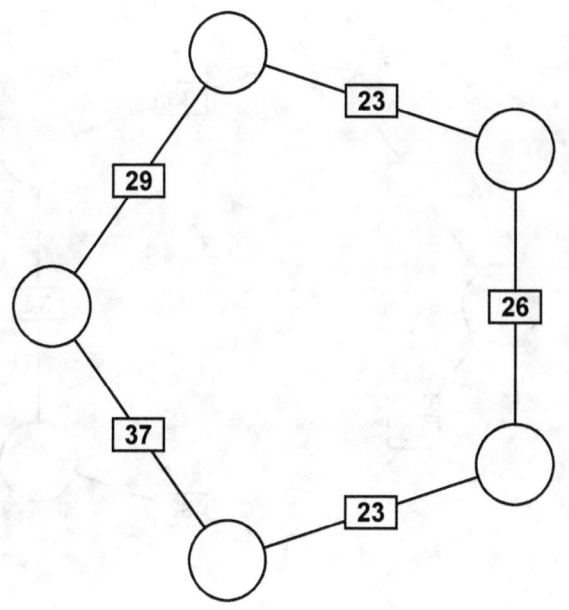

Puzzle 356

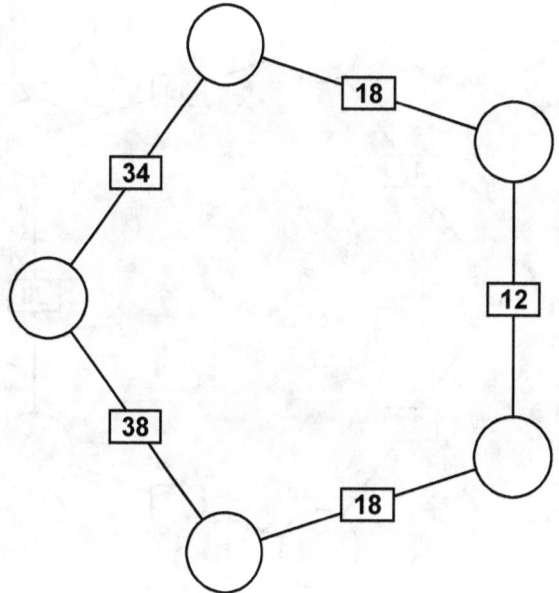

Puzzle 357

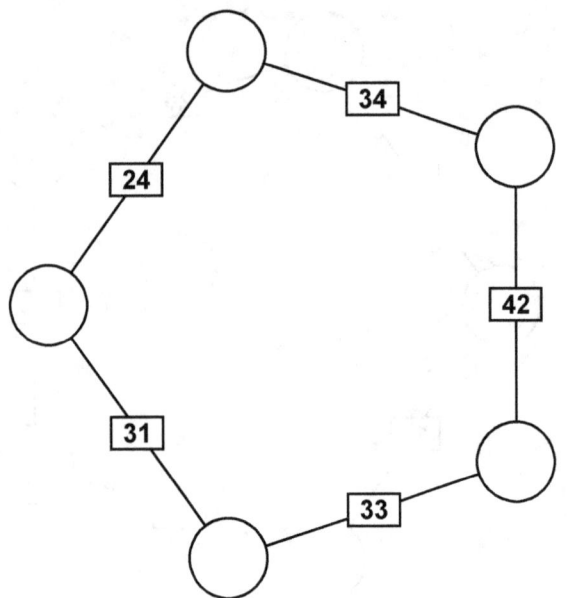

Puzzle 358

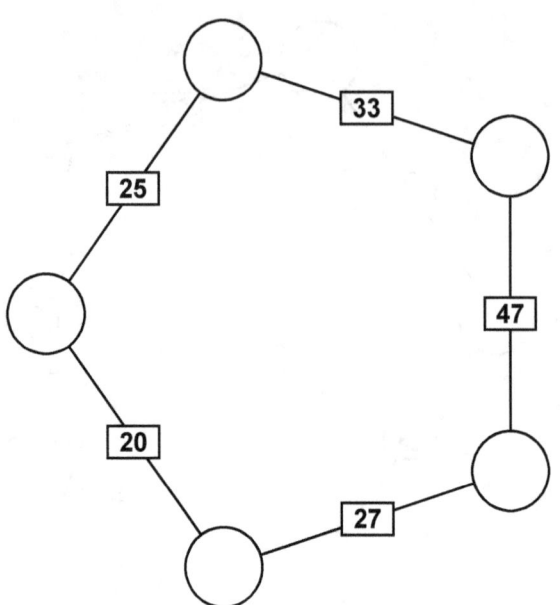

Puzzle 359

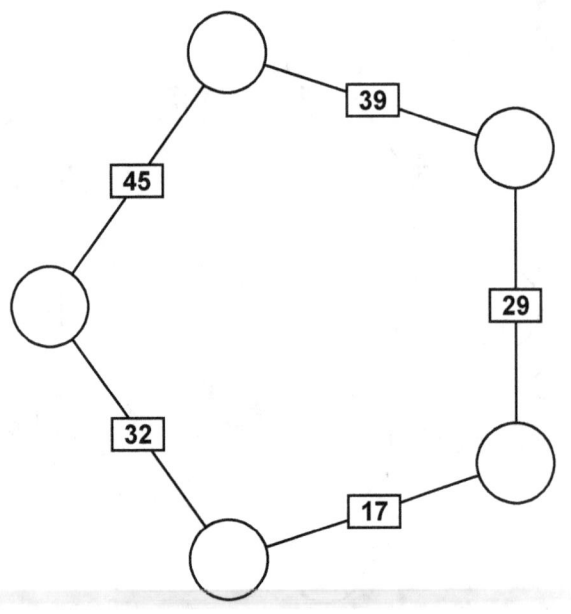

Puzzle 360

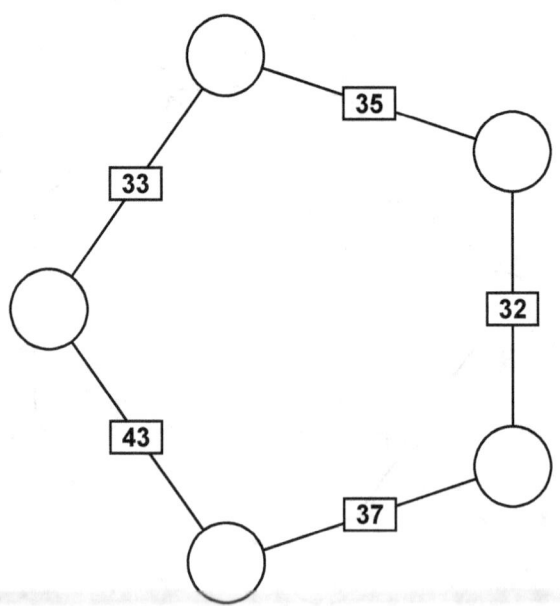

Puzzle 361

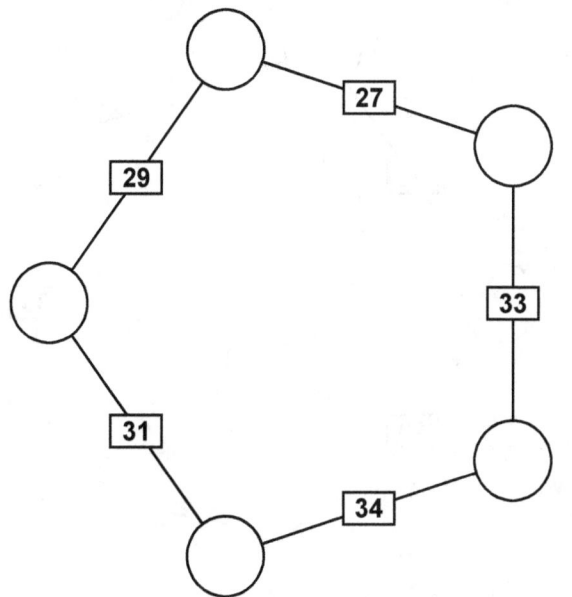

Puzzle 362

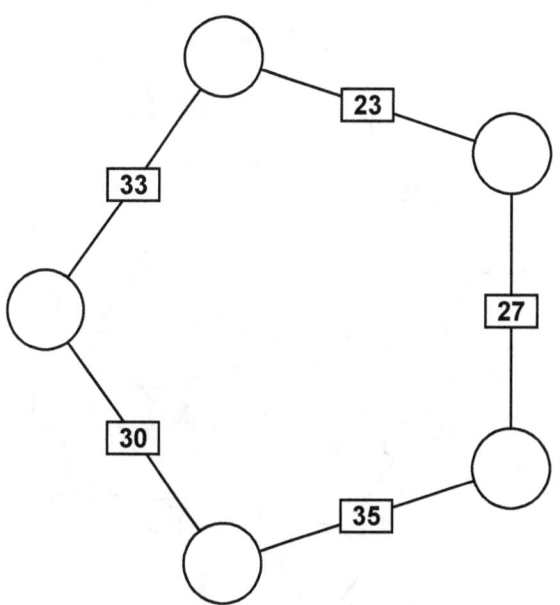

Puzzle 363

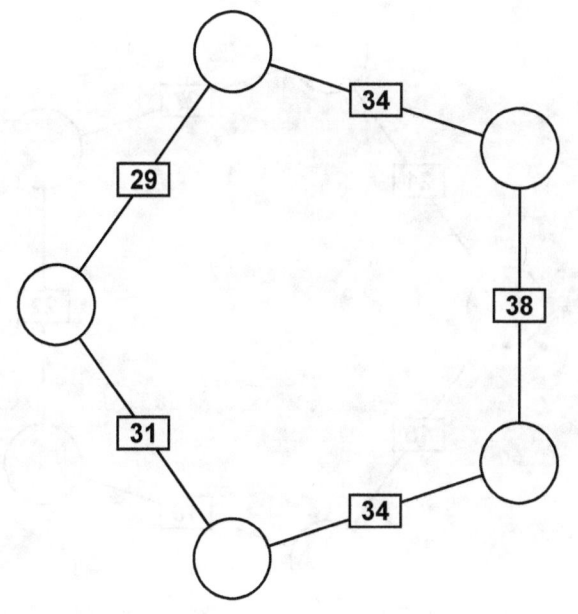

Puzzle 364

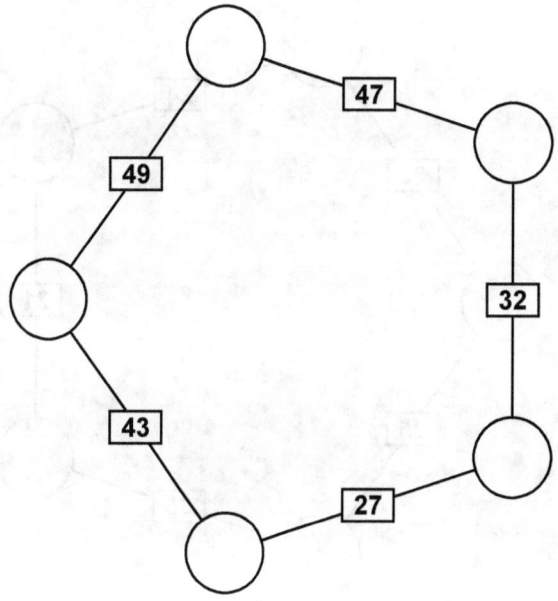

Puzzle 365

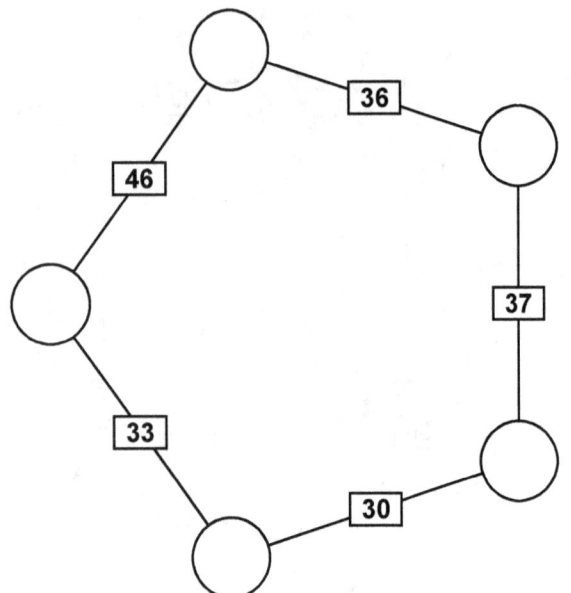

Puzzle 366

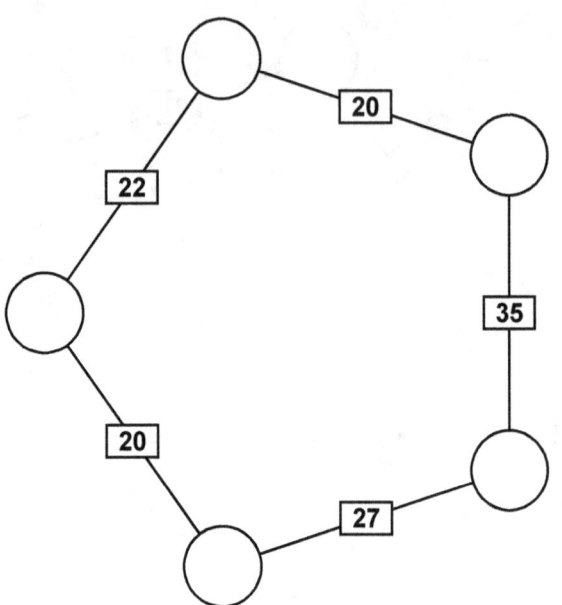

Puzzle 367

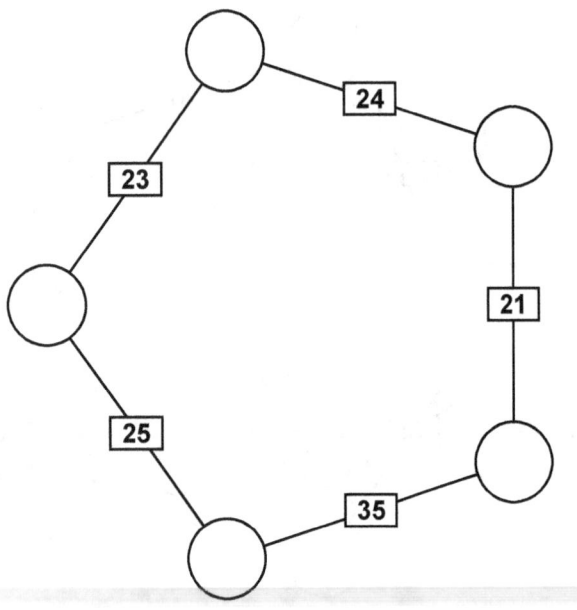

Puzzle 368

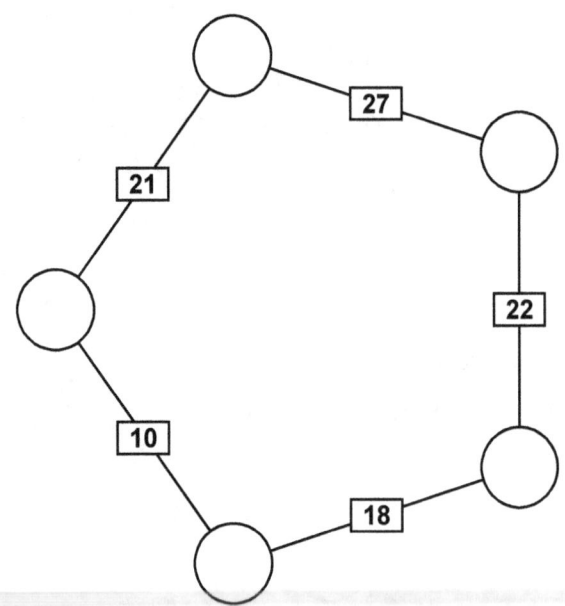

Puzzle 369

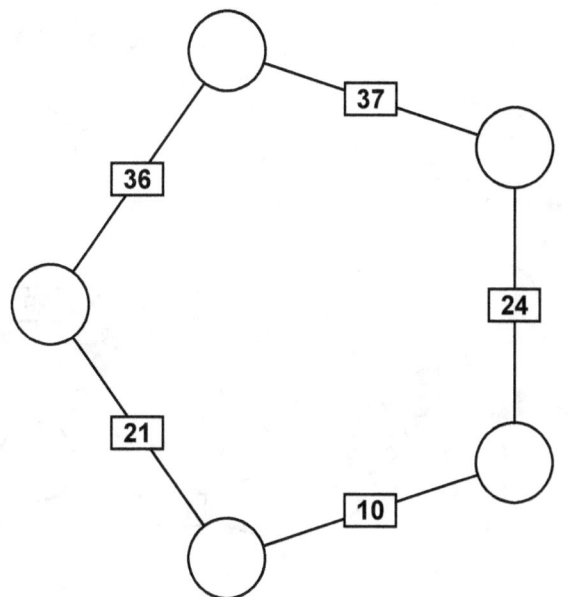

Puzzle 370

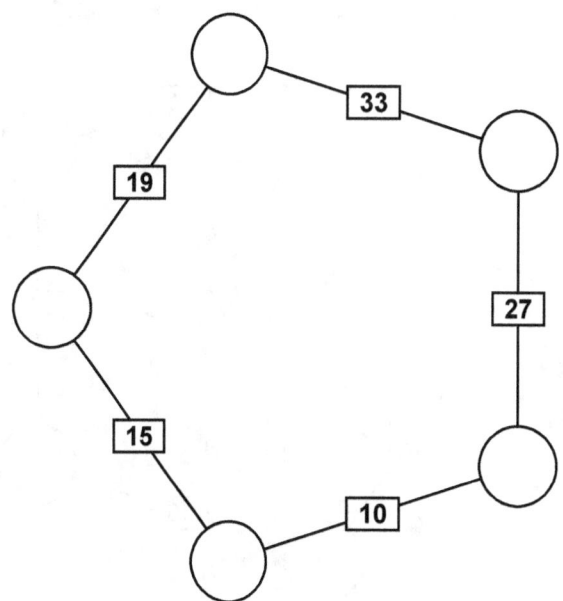

Puzzle 371

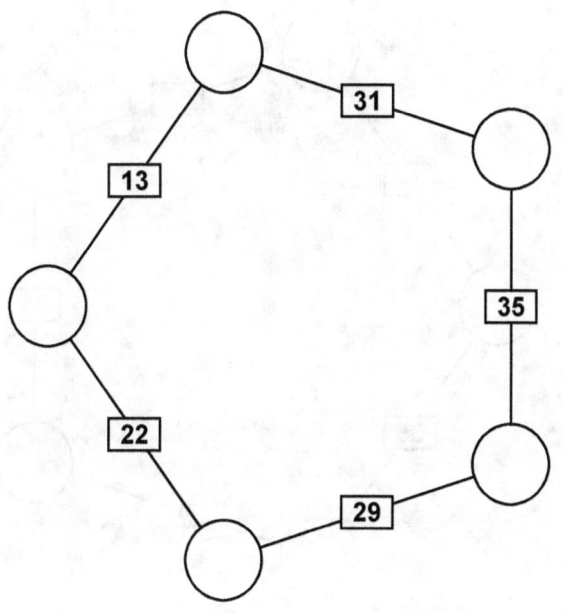

Puzzle 372

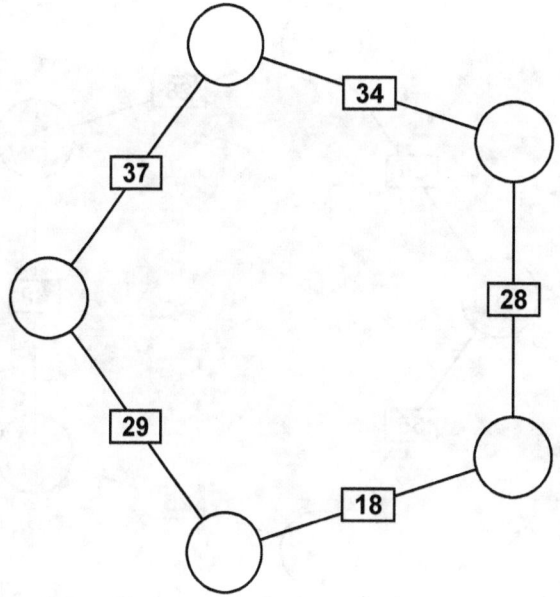

Puzzle 373

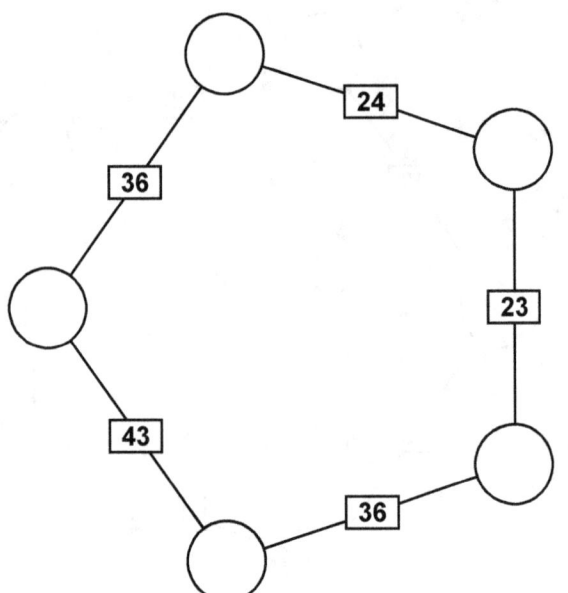

Puzzle 374

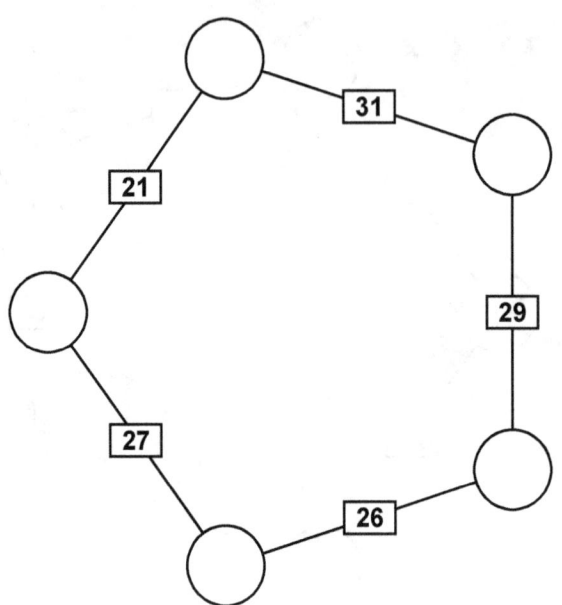

Puzzle 375

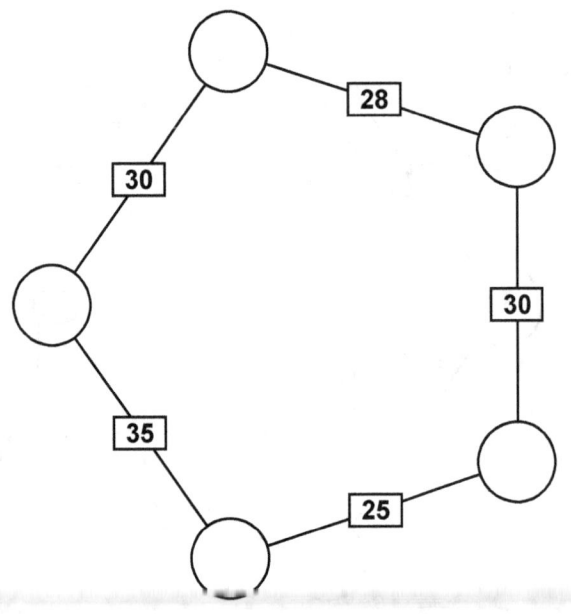

Puzzle 376

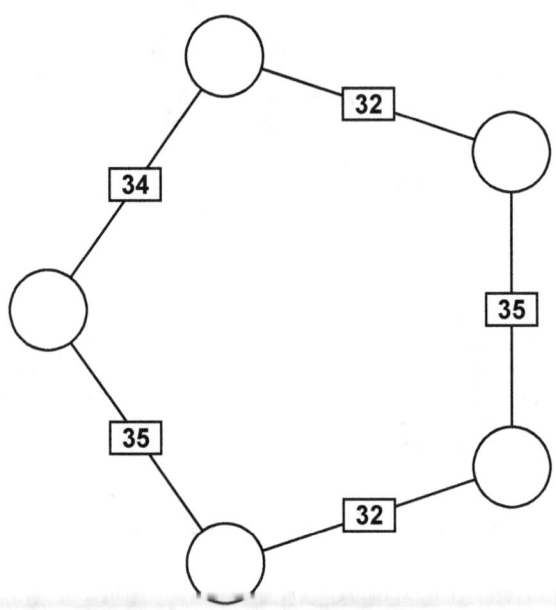

Puzzle 377

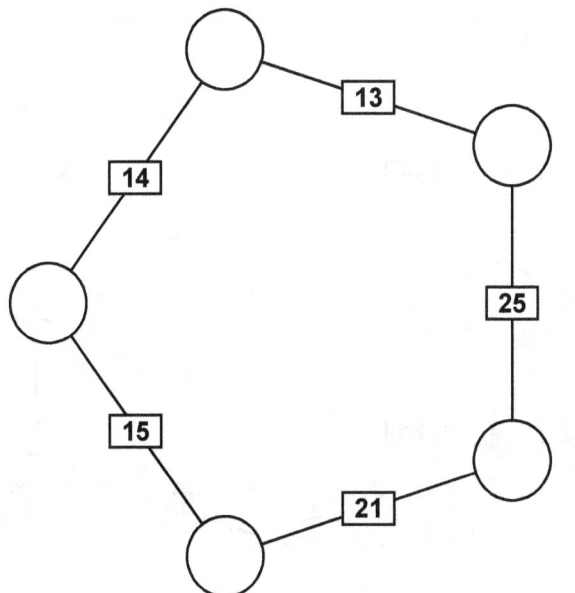

Puzzle 378

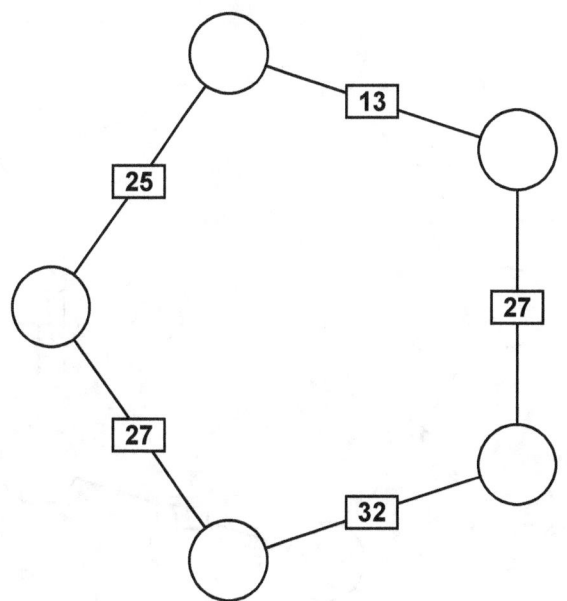

Puzzle 379

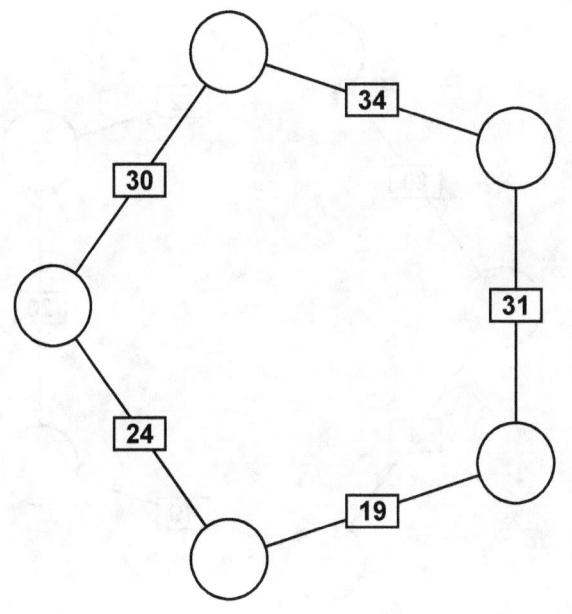

Puzzle 380

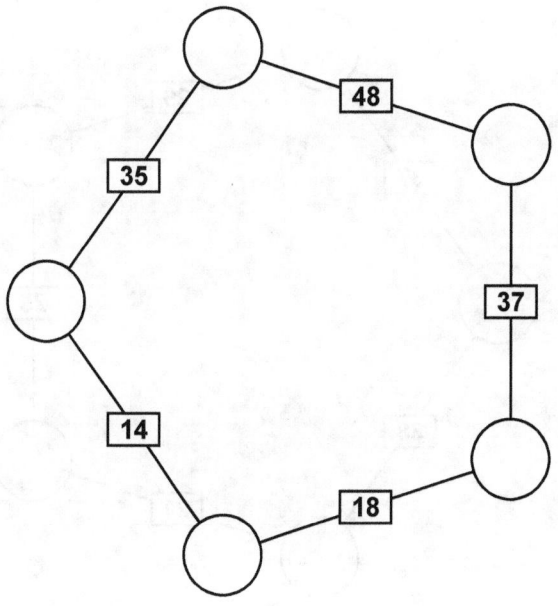

Puzzle 381

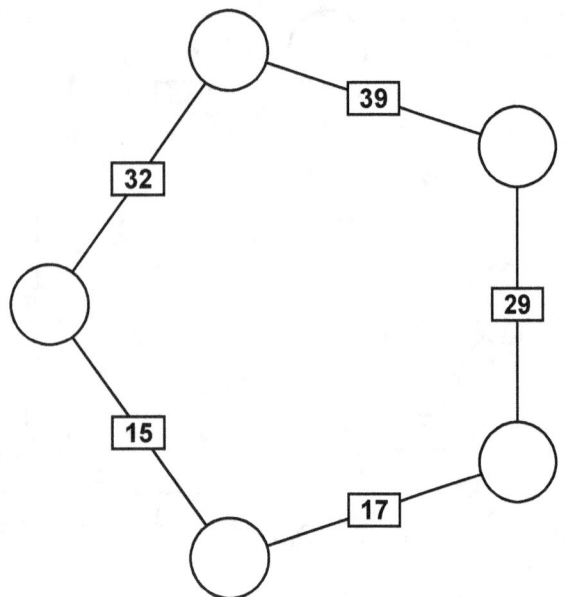

Puzzle 382

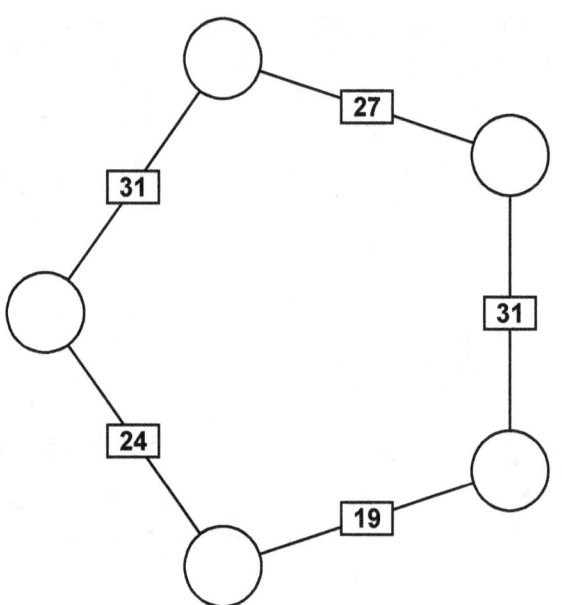

Puzzle 383

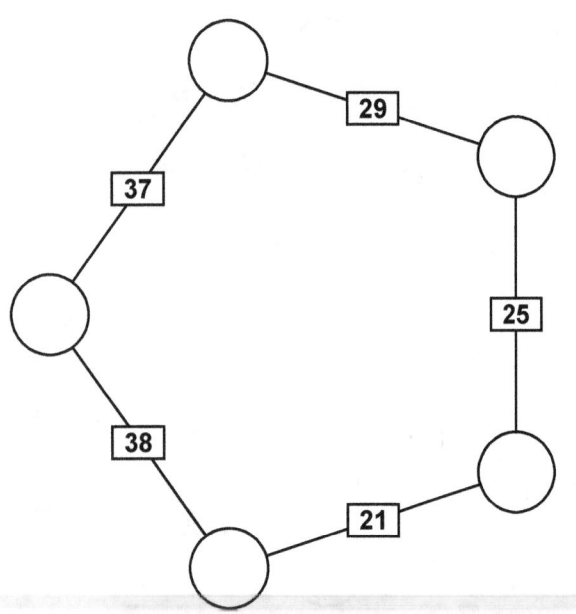

Puzzle 384

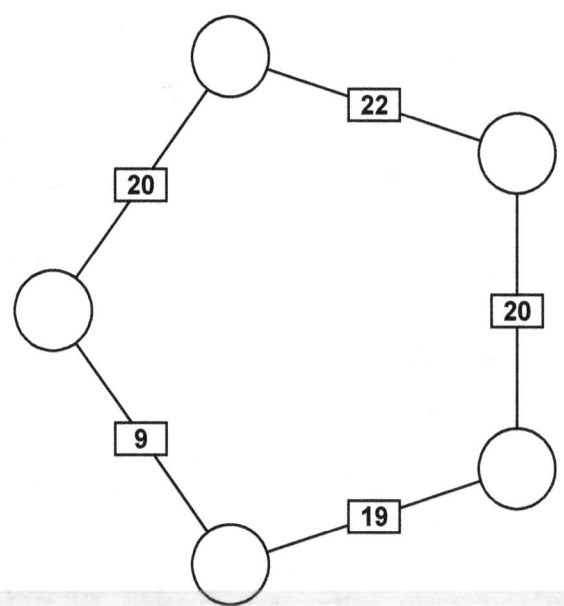

Puzzle 385

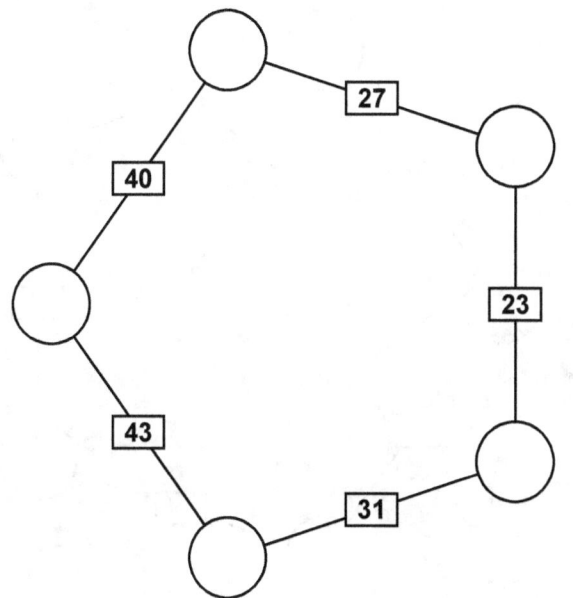

Puzzle 386

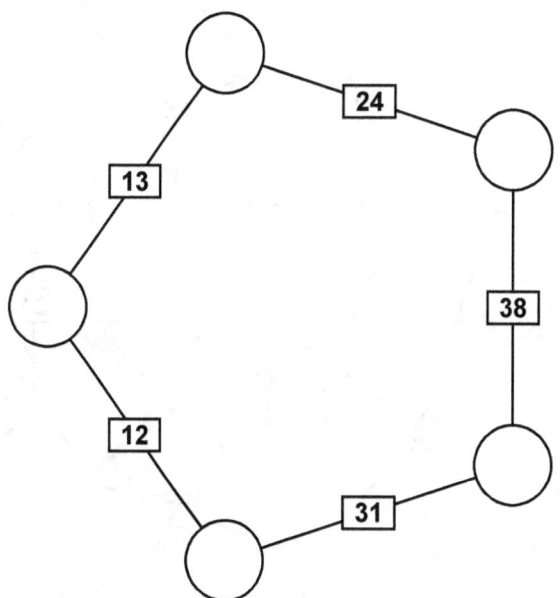

Puzzle 387

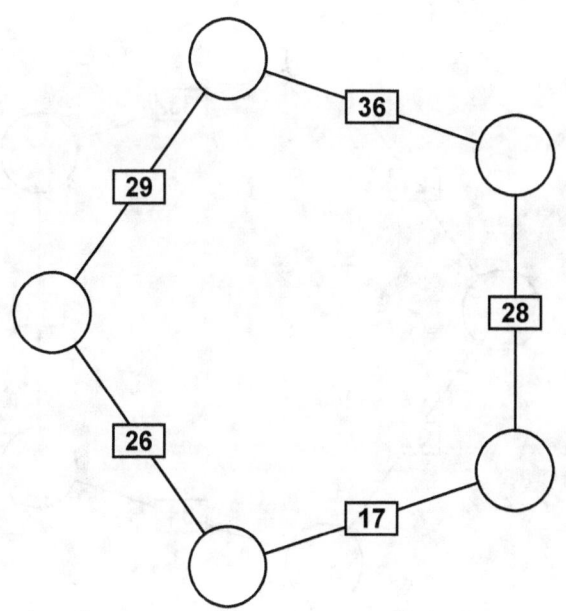

Puzzle 388

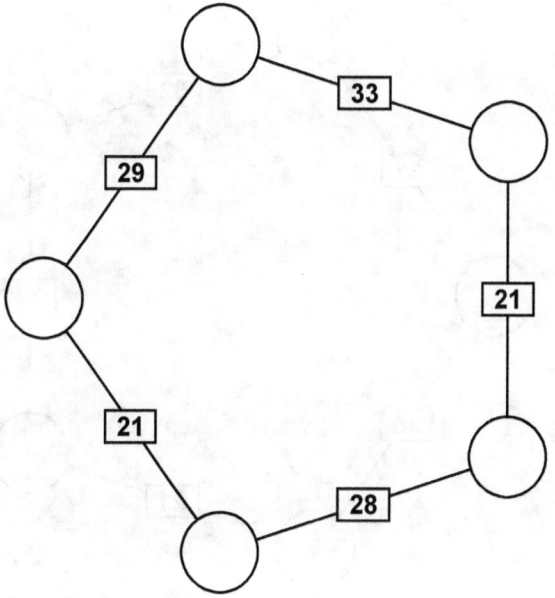

Puzzle 389

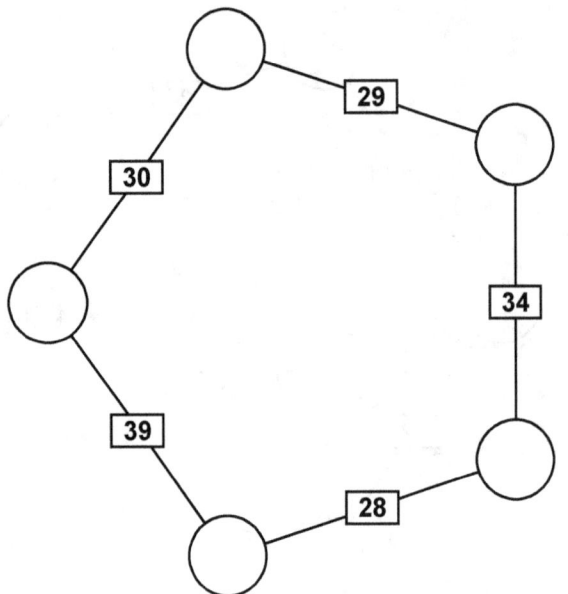

Puzzle 390

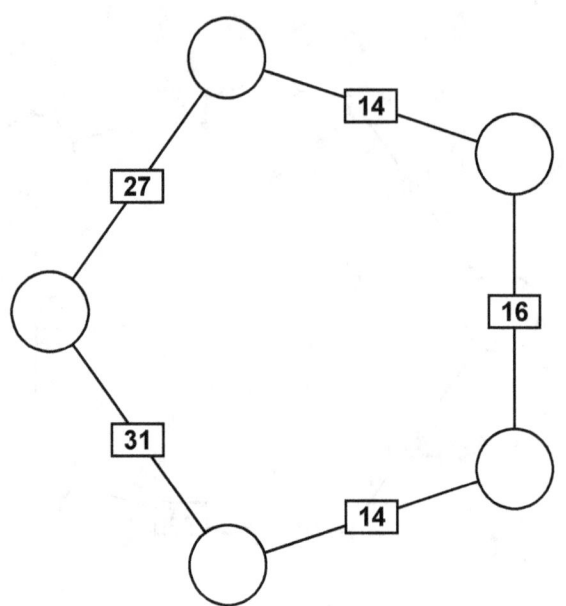

Puzzle 391

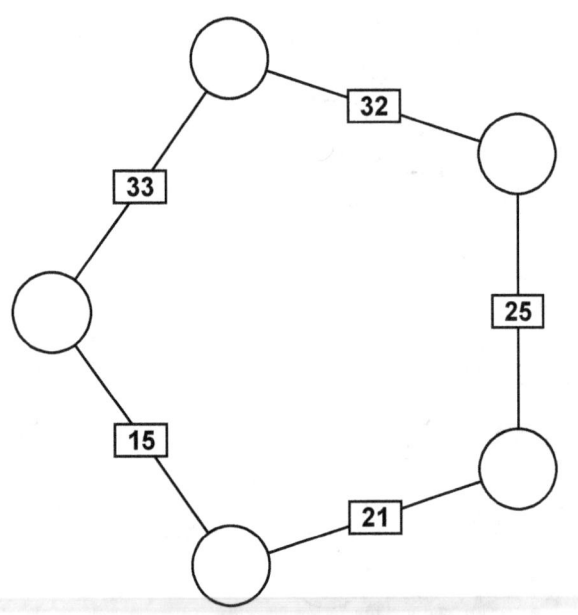

Puzzle 392

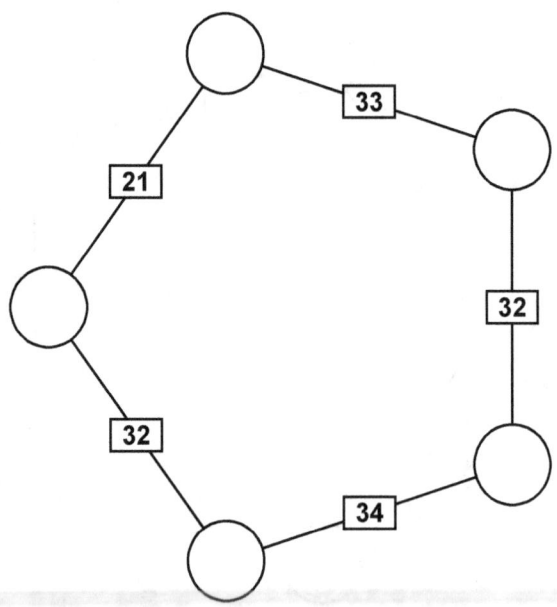

Puzzle 393

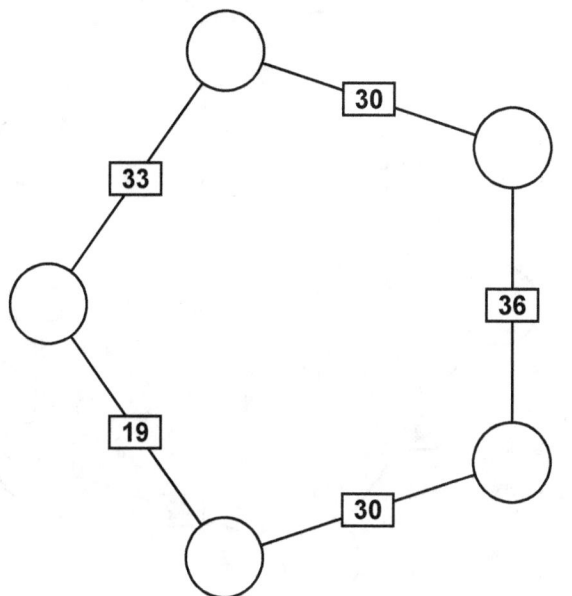

Puzzle 394

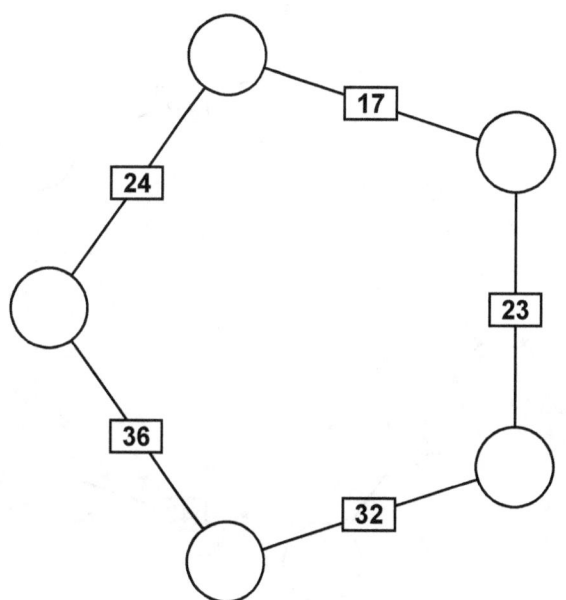

Puzzle 395

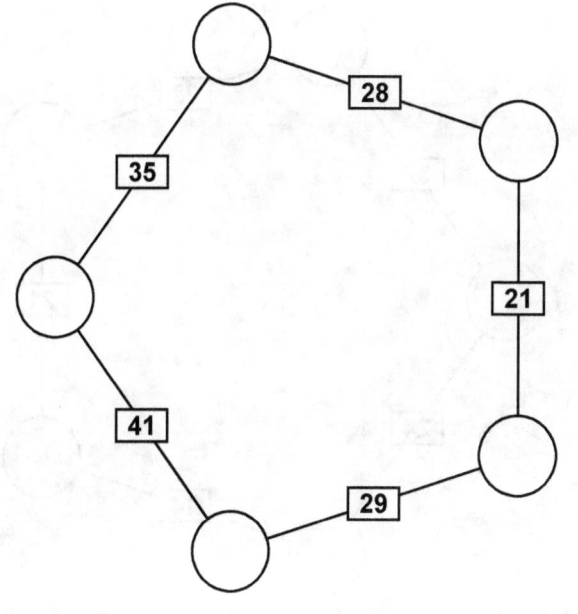

Puzzle 396

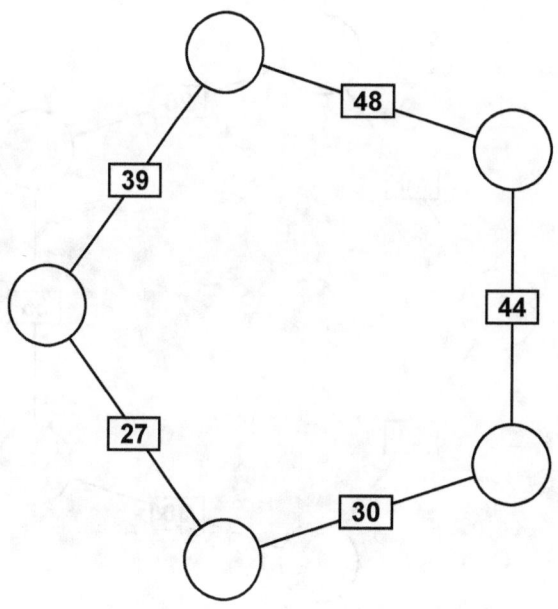

Puzzle 397

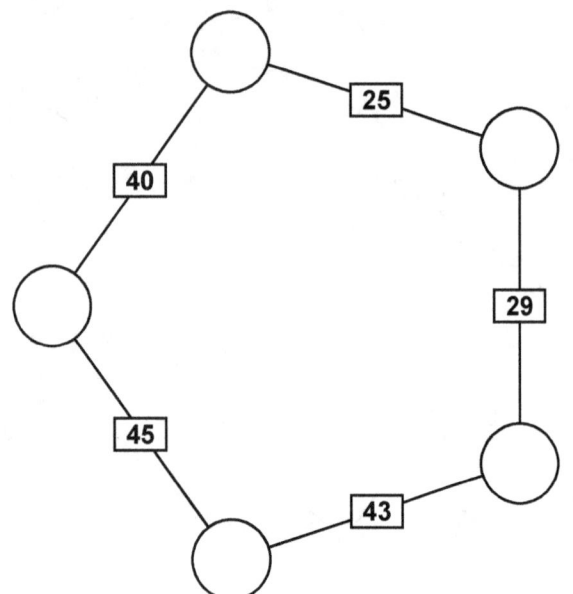

Puzzle 398

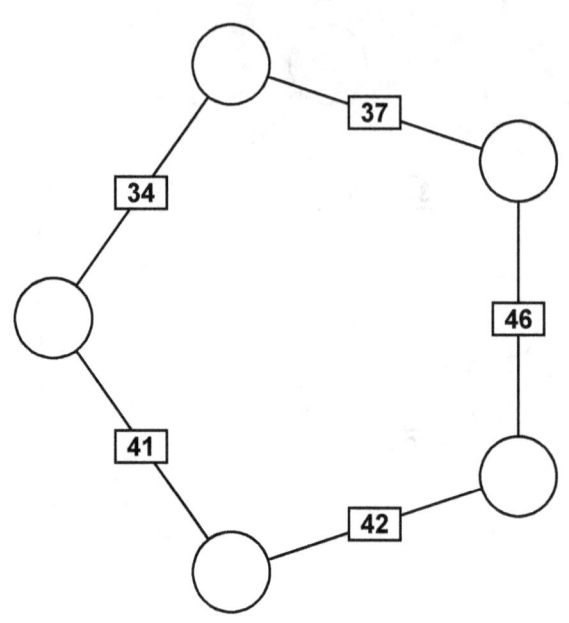

Puzzle 399

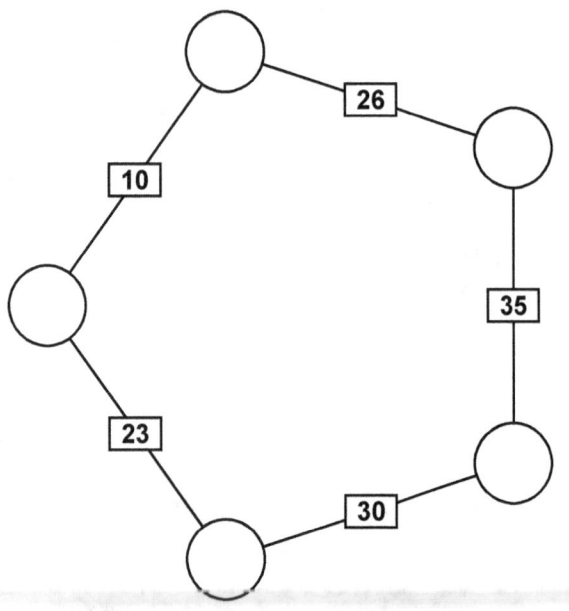

Puzzle 400

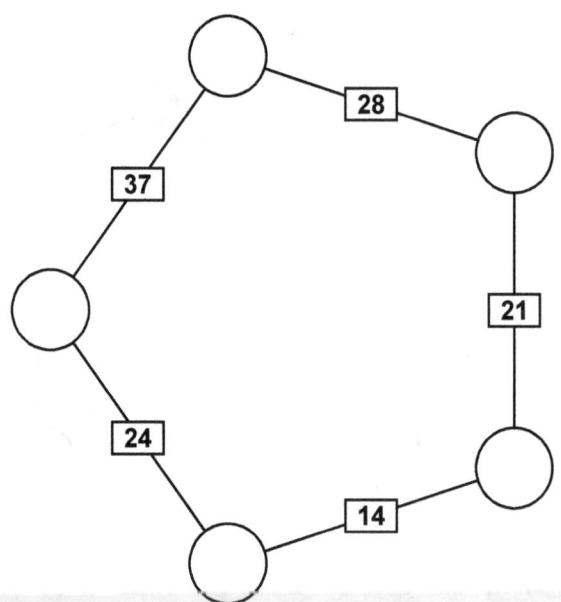

www.ingramcontent.com/pod-product-compliance
Lightning Source LLC
Chambersburg PA
CBHW081200180526
45170CB00006B/2164